HOW TO FEED
THE WORLD

By the same author

China's Energy
Energy in the Developing World (*edited with W. Knowland*)
Energy Analysis in Agriculture (*with P. Nachman and T. V. Long II*)
Biomass Energies
The Bad Earth
Carbon Nitrogen Sulfur
Energy Food Environment
Energy in China's Modernization
General Energetics
China's Environmental Crisis
Global Ecology
Energy in World History
Cycles of Life
Energies
Feeding the World
Enriching the Earth
The Earth's Biosphere
Energy at the Crossroads
China's Past, China's Future
Creating the 20th Century
Transforming the 20th Century
Energy: A Beginner's Guide
Oil: A Beginner's Guide
Energy in Nature and Society
Global Catastrophes and Trends
Why America Is Not a New Rome
Energy Transitions
Energy Myths and Realities
Prime Movers of Globalization
Japan's Dietary Transition and Its Impacts (*with K. Kobayashi*)
Should We Eat Meat?
Harvesting the Biosphere
Made in the USA
Making the Modern World
Power Density
Natural Gas
Still the Iron Age
Energy Transitions (new edition)
Energy: A Beginner's Guide (new edition)
Energy and Civilization: A History
Oil: A Beginner's Guide (new edition)
Growth
Numbers Don't Lie
Grand Transitions
How the World Really Works
Inventions and Innovations
Size

HOW TO FEED THE WORLD

The History and Future of Food

VACLAV SMIL

VIKING

VIKING
An imprint of Penguin Random House LLC
1745 Broadway, New York, NY 10019
penguinrandomhouse.com

Copyright © 2024 by Vaclav Smil

Penguin Random House values and supports copyright. Copyright fuels creativity, encourages diverse voices, promotes free speech, and creates a vibrant culture. Thank you for buying an authorized edition of this book and for complying with copyright laws by not reproducing, scanning, or distributing any part of it in any form without permission. You are supporting writers and allowing Penguin Random House to continue to publish books for every reader. Please note that no part of this book may be used or reproduced in any manner for the purpose of training artificial intelligence technologies or systems.

VIKING is a registered trademark of Penguin Random House LLC.

Image credits: p. 6, Ronan Donovan; pp. 11, 52, 78, 90, private collection; p. 31 (top), Shutterstock; p. 31 (bottom left), Wikipedia; p. 31 (bottom right), Pixabay; pp. 35, 65, 164, 166, Alamy; p. 87, 2014 Poultry Science Association Inc.

Set in Bembo Book MT Pro

LIBRARY OF CONGRESS CONTROL NUMBER: 2024952420

ISBN 9780593834510 (hardcover)
ISBN 9780593834527 (ebook)

First published in hardcover in Great Britain by Viking, part of the Penguin Random House Group of Companies, Penguin Random House Ltd., London, in 2024

First United States edition published by Viking, 2025

Printed in the United States of America
1st Printing

The authorized representative in the EU for product safety and compliance is Penguin Random House Ireland, Morrison Chambers, 32 Nassau Street, Dublin D02 YH68, Ireland, https://eu-contact.penguin.ie.

Contents

Acknowledgments	vii
Introduction	1
1. What Did Agriculture Ever Do for Us?	5
2. Why Do We Eat Lots of Some Plants and Not Others?	29
3. The Limit of What We Can Grow	55
4. Why Do We Eat Some Animals and Not Others?	77
5. What's More Important: Food or Smartphones?	101
6. What Should You Eat to Be Healthy?	125
7. Feeding a Growing Population with Reduced Environmental Impacts: Dubious Solutions	147
8. Feeding a Growing Population: What Would Work	173
References and Notes	203
Index	249

Acknowledgments

As with any interdisciplinary book, this review and analysis of food and the global food system could not have been written without relying on hundreds of scientists whose publications, recent and classical, helped me to understand the scope of our accomplishments and constraints. And special thanks to Connor Brown, my London editor, who indulges my preference for writing books on very different topics; to Gemma Wain, for another careful edit; to Kenneth Cassman, Emeritus Robert B. Daugherty Professor of Agronomy at the University of Nebraska, for his critical reading of the typescript and for his corrections and suggestions; and to Bill Gates, who has been my regular reader and critical reviewer for more than 15 years and who suggested, sometime in the second pandemic year, that I should take another look at food.

Introduction

Catastrophism has a long historical pedigree, and worries about feeding the world have been with us ever since Thomas Robert Malthus published his *Essay on the Principle of Population* in 1798, with its warning that "the power of population is indefinitely greater than the power in the earth to produce subsistence for man."

Thus was born the idea that the human population grows more rapidly than the food supply, until checks on that growth—famine, war, or disease—reduce the population; and therefore, over time, the population remains stagnant.

But it is telling of the extent of half-remembered truths in the history and science of food production that on a closer look, even Malthus was not a "Malthusian" after all. The second (1803) edition reveals him to be more optimistic: "though our future prospects . . . may not be so bright as we could wish, yet they are far from being entirely disheartening, and by no means preclude that gradual and progressive improvement in human satiety." Sadly, facts are of little concern to those who sell agendas.

As the global population continues to grow and as environmental worries increase, concerns about feeding the world remain, and some of them are rather dire. For example, in the *Guardian* in May 2022, British writer and political activist George Monbiot claimed that "the global food system is beginning to look like the global financial system in the run-up to 2008. While financial collapse would have been devastating to human welfare, food system collapse doesn't bear thinking about. Yet the evidence that something is going badly wrong has been escalating rapidly."

This is just one example in a sea of dubious claims and outright misinformation. Over the past decade I have been repeatedly exasperated by people's poor understanding and sheer ignorance of life's many basic realities, be they concerning organisms or machines, crops or engines, food or fuels.

So should you be worried about the global food system? Do you live in a place that will be blighted by famine in the coming decades? Is society going to collapse? The short answer is: probably not. A fuller answer, which draws on the history of food production and the latest scientific understanding, explaining some key biophysical factors like photosynthetic efficiency and nutrient needs, is longer—about the length of this book.

If you are looking for a text about stunningly disruptive innovations that will soon revolutionize the food system, this is not that book. This is the opposite: it argues for the power of incremental changes, the sorts of things often ignored by the media and writers of popular non-fiction, who instead focus on the unrealistic. Moreover, I do not understand the need for hyperbolic and incorrect statements when the actual numbers are newsworthy and attention-grabbing enough.

For example, global food production now averages about 3,000 kcal per person; daily global food waste is about 1,000 kcal per person. And yet there is no urgency about changing that. If you were, constantly, losing a third of your income, you'd try to do something about it. This book looks at such realities.

Why have we domesticated such a small number of plants and animals to produce food? If our ancient ancestors had today's evidence at hand, would they choose differently? What do the best available studies say about the latest diet fads, from keto to avoiding ultra-processed foods? Looking ahead to 2050, will the world have freed its livestock and be living in a techno-vegan utopia, fueling our guilt-free existence with plant-based or lab-grown meat substitutes? I am a proponent of reducing our meat intake—a third of global cereal production and two-thirds of the US grain harvest is fed to animals—but if doing so means eating more fruits and nuts, for example, then it may not be any better for the environment.

What about organic farming? Is that the panacea? In previous centuries where the available technology meant that all farming was "organic," commonly 80 percent of people worked in farming, doing unglamorous jobs like collecting manure to use as fertilizer. In rich countries today, no more than 1–3 percent of people produce food. Would you like to collect manure?

More importantly, the very notion of agriculture as the fundamental existential endeavor has been under attack. Did the advent of farming allow human flourishing or—as has been argued by many writers of popular non-fiction—was it the greatest catastrophe in history? In this book, we will critically assess the alternatives.

How to Feed the World marks the beginning of the fifth decade of my work on food. This began in the late 1970s with research for a more specialist book—the first book-length energy analysis of corn, America's leading crop (published in 1982)—and five of my books that appeared during the 1980s had segments or chapters devoted to crops and food.

In 2000, I published my first book devoted solely to many aspects of food: *Feeding the World*, which covered topics from photosynthesis and crop yields to animal husbandry and diets. This was immediately followed in 2001 by *Enriching the Earth*, a detailed account of modern agriculture's most fundamental input: ammonia, used—as we will see later—in the production of all nitrogenous fertilizers. Three of my books published during the first decade of the twenty-first century returned to food: *Japan's Dietary Transition and Its Impacts* (with Kazuhiko Kobayashi), *Harvesting the Biosphere* and *Should We Eat Meat?*

Since 2014 I have worked on other topics, publishing books on steel, oil, natural gas, energy transitions, energy and civilization, growth and size, although *How the World Really Works* (2022) has a chapter on understanding food production. Simply put, food is not a passing interest of mine.

That said, food and agriculture are subjects with enormous factual and intellectual scope, and hence every broader review must take place within self-imposed limits. And so this book explains the basic properties of the global food system. To do that, I take a quantitative approach, because when it comes to food, numbers are much more important than opinions and feelings. We will look at everything from agronomy and crop science to energy accounting and nutrition and health, and will do so by following a logical sequence of eight essential topics.

The book's first half is devoted to the biophysical basics of growing food. In the second half I quantify the real scope of the global food system, explain dietary necessities, and cast a critical eye over some recent suggestions for radical transformation of this system. Readers expecting extended coverage or critiques of two fashionable topics—agriculture and global climate change, and sustainable farming—should look elsewhere. This is not yet another book about food and global warming: so many books have been published on this sprawling topic that you could now own a small library of such writings.

Deliberately, this book is not intended as a comprehensive review of modern food production and nutrition, but as a focused, strongly quantitative evaluation of the basics. Many books about agriculture and food do not contain many numbers, but this book is teeming with them. I am unapologetic about this. Numbers are the antidote to wishful thinking and are the only way to get a solid grasp of the modalities and limits of modern crop cultivation, food, and nutrition. With this foundation it is far less likely you will make incorrect interpretations or misunderstand the basic realities of food, nor will you accept uncritically the many exaggerated claims and unrealistic promises regarding the future of global farming.

1. What Did Agriculture Ever Do for Us?

Why do we need agriculture? Why do we have to cultivate annual and perennial crops? Why does cropland cover nearly 40 percent of the planet's ice-free land? Why do we keep billions of domestic animals? The answer to all these questions is: because there are so many of us. And as is often the case with increasing quantities, the result is a fundamental change in quality.

Our species separated from primates more than 6 million years ago, and then evolution led, about 300,000 years ago, to the emergence of *Homo sapiens*—us. As long as our ancestors lived in small and widely dispersed groups, they could survive in the same way their primate predecessors did—as gatherers and hunters. While the diets of these hominin species cannot be reconstructed in detailed quantitative terms (the best tools we have available, such as analyses of stable isotopes in preserved bones and teeth, cannot give such detailed answers), chimpanzee foraging provides a realistic template for their qualitative recreation. From this we can infer that hominins ate a wide variety of plants and small animals through opportunistic scavenging, the deliberate hunting of smaller prey, and occasionally even cannibalism.[1]

The chimp diet

Numerous studies document the omnivorous eating habits of chimpanzee groups in tropical Africa: the large variety of species they eat, their preference for readily digestible plant matter, their consumption of insects and their hunting of small mammals.[2] Forest chimpanzees commonly eat more than 100 different plant species, but fruits dominate their diet (with figs being their favorite),

Our omnivorous predecessors: chimpanzees killing and eating monkeys.

supplemented by flowers, fresh leaves and stems, pith, roots, seeds, and nuts, some of which they crack by using small stone hammers. Many field observations have detailed how chimpanzees search for insects (above all for termites, often by "fishing" with grass blades) and for invertebrates, bird eggs, and chicks.

Chimpanzees also hunt small mammals (mostly colobus monkeys but also young wild pigs, bushbucks, bush babies, blue duikers, and baboons), and subsequently share their meat with others in the group. In Gombe, Tanzania, adult male chimpanzees have been recorded hunting as much as 25 kilograms of such meat per year; substantially more than is eaten in most traditional agricultural societies, where annual per capita meat consumption was less than 10 kilograms. The hunting of small animals is done mostly by two or more males with encouraging success rates of 50–60 percent, but females also hunt, even when carrying young. And at the Fongoli study site in Senegal, chimpanzees were observed to use many kinds of spear-like tools to kill bush babies—small nocturnal prosimians that sleep in tree cavities during the day.[3] The risks of hunting (injuries resulting from fast chases through canopies; prey resistance) are well rewarded: after all, even a small bite of meat provides more nutrition (above all, more protein) than do hundreds of termites, which take a long time to capture.

In tropical forest environments with an abundance of plant and animal species, this is not a highly taxing existence and forest chimpanzees spend roughly half of the daylight hours looking for food and eating, with between 60 and 80 percent of their feeding time devoted to searching for and eating fruit. That leaves plenty of time for resting, exploring, socializing, and grooming. But an omnivorous diet high in fruits limits the number of individuals in a group (and hence their maximum density within an exploited area) because there are only so many fruit trees to harvest, most of them produce only one or two crops a year, and other species compete for this limited production. Some forested environments can support an average of 1.5 chimpanzees per square kilometer, or up to two or even four individuals in the best fruit-providing areas, while in open, and often degraded and arid, savannah environments, typical densities are less than one individual per 2 square kilometers.[4] Living off wild fruits and the small animals that you and your family have captured and killed is clearly impossible in today's high-density urban environment.

Hominin and early human diets

The diet of the early hominins who diverged from chimpanzees more than 6 million years ago continued to resemble the omnivorous pattern described above. Food intake was dominated by the consumption of plant tissues (fruits, tubers, nuts, leaves), which are easily digestible and can provide the necessary nutrition. This was supplemented by the moderate consumption of invertebrates and small vertebrates, and by opportunistic meat and marrow scavenging from kills made by large carnivores.[5] Later advances in tool-making, starting with small stone tools and eventually progressing to spears and bows and arrows, made it possible to hunt and butcher larger animals.

Modern anthropological evidence overwhelmingly shows that the human position on the food chain evolved from the comparatively lower chimpanzee-level of meat-eating, to a high level of carnivory that peaked in *Homo erectus* (a species that survived until about 250,000 years ago), and began to reverse in the Upper Paleolithic (or late Stone Age), some 50,000–12,000 years ago.[6] The evidence, which can be observed in human remains found across the world, includes higher fat reserves and higher stomach acidity over time, a changing gut shape and volume (which limited the ability to extract energy from plant fibers), the reduction of chewing muscles (less mastication was required with better diets), and earlier weaning (as milk was supplemented and then replaced by more nutritious food).

In colder climates the dietary shift was affected by the extinction of the largest terrestrial mammals—the megaherbivores, such as mammoths—that took place during the Neolithic (or New Stone Age), 9,000–3,000 BCE. The two competing hypotheses explaining that demise have been climate change, resulting in the expansion of forests and retreat of grasslands, where these huge animals lived, and (a much less likely but persistent and hugely popular explanation) overkill—extinction caused by the mass-scale killing of large herbivores by groups of prehistoric hunters.[7]

Although *Homo sapiens* eventually expanded the scope of its foraging skills to the killing of megaherbivores and to fishing in fresh and coastal waters (and hence was able to survive in environments ranging from the tropics to the Arctic), population densities of gatherer-hunter groups remained limited. Patchy archeological records mean it is impossible to make any reliable reconstructions of actual prehistoric densities, but we have a great deal of reliable quantitative information concerning the numbers and ways of acquiring food of those foragers who survived into the 20th century, to be studied by anthropologists.[8] As American anthropologist Frank Marlowe, who studied the Hadza hunters and gatherers of

Tanzania, noted, while these "foragers may be problematic analogs of humans in the past, they are certainly the most useful exemplars of humans in the present"—and the wide range of encountered group sizes and population densities (in Southern and Central Africa, in the Amazon, and in Australia) is likely to encompass most of the experiences of the earlier groups that relied on a limited range of simple tools.[9]

From these studies, we can see that the smallest group of foragers required for survival appears to be 25–30 people, while the largest sizes of sedentary groups of fishers, hunters, and gatherers cluster around 500. Across 300 studied foraging societies that persisted into the 19th and 20th centuries, the mean population density was 0.25 persons per square kilometer. The smallest was below 0.1, while the largest—above one person per square kilometer—were exceptions reached by sedentary societies with access to highly nutritious (and fatty) marine foods such as fish and seals. For example, large groups of around 500 people in the Pacific Northwest were dependent on easily caught salmon during the annual migration (and some even hunted small whales in coastal waters).[10] Very few settings offered such abundant food supply that they were able to support large sedentary communities.

When average body weights are considered (an adult human woman is 55 kilograms, an adult chimpanzee female is 35 kilograms), the density range of human foragers has a remarkably close (but not surprising) overlap with the density range of chimpanzees: on average, environments would have been able to nourish between 5 and 50 kilograms of live body mass per square kilometer. The lowest population densities of land-based foragers were among sub-Arctic and other high-latitude groups, as well as in dry savannah environments, but there was a relatively wide range of population densities even in more hospitable circumstances, for example in seasonally dry Mediterranean climates and in tropical rainforests. These limited density ranges show clear limits to gathering and hunting in terms of the energy that could be harvested, be it by primate quadrupeds or by bipedal humans; so, even when we began to consume the meat of larger animals, hunting and gathering could never support very large individual groups (in this context, this means thousands of people, never mind the tens of millions in the largest cities of today), nor relatively high densities exceeding 10 people per square kilometer (today, Manila in the Philippines has a population density two orders of magnitude greater, at more than 70,000 people per square kilometer).

Our growing population

Recent genetic studies and demographic models mean we are surer of prehistoric population totals than we have ever been.[11] The number of human ancestors who lived more than 1.2 million years ago may not have been higher than about 20,000 hominin individuals, which is far fewer than the global counts of chimpanzees and gorillas today. Later hominin populations (like *Homo erectus* and *Homo heidelbergensis*) likely rose from only 50,000 a quarter of a million years ago, to 100,000 individuals of *Homo sapiens*—us—some 100,000 years ago. Genetic evidence indicates that subsequent, and relatively substantial, population growth was followed by a sudden retreat brought on by severe planetary cooling and the expansion of the ice sheets between 29,000 and 17,000 years ago.

In 2015, a group of Finnish scientists concluded that the European *Homo sapiens* population declined from more than 300,000 people 30,000 years ago to about 130,000 individuals 23,000 years ago, and then grew to about 400,000 by the end of the last ice age, about 10,000 years ago.[12] At the beginning of the Neolithic, some 12,000 years ago, humans lived in environments ranging from tropical rainforests to the Arctic. In the rainforests, where large animals are rare and small animals live mostly in the trees, are often nocturnal, and are always difficult to kill, diets had to be dominated by plants.[13] In temperate latitudes, the meat of large herbivores supplied substantial shares of total energy; in the Arctic, survival was impossible without killing large fatty marine mammals.[14]

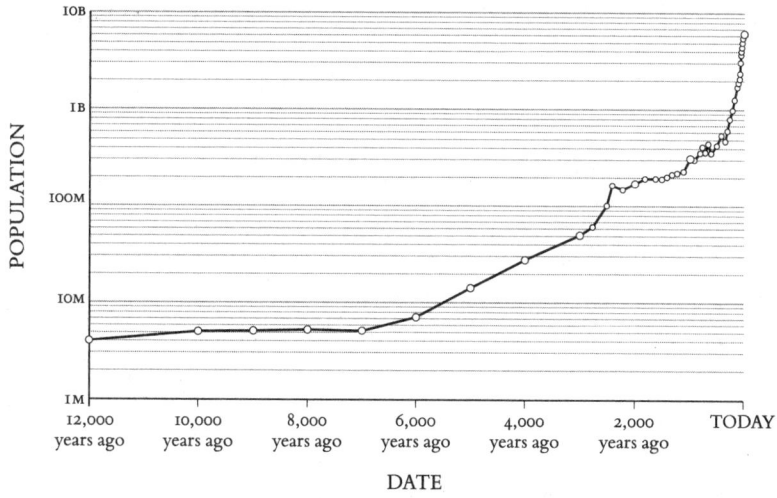

World population growth: prehistoric stagnation followed by a slow take-off.

Of course, the reader knows *something* happened to facilitate population growth. No prizes for guessing—especially given the chapter title—that this was due to the subsequent domestication of crops and animals, which made it possible to support far higher population densities. But, contrary to the much-repeated claim introduced by Gordon Childe, an Australian archeologist working in the UK, in his influential book *Man Makes Himself* (1936), that this amounted to a so-called Neolithic revolution, it was a gradual development, a millennia-long process during which increasing reliance on planted crops was augmented by the gathering of wild plant species and the killing of wild animals; indeed, in many places, these foraging activities supplied non-negligible shares of food energy in long-established agricultural societies, and even today scores of wild foods are harvested by settled populations in many African and Asian countries.[15] Before the advent of domestication (this began about 12,000 years ago in several regions of the Middle East), the most likely range of global population was 2–4 million people; by the beginning of the Common Era (that is when Augustus was the first Roman emperor, around 2,000 years ago), the most likely range of the global population was roughly a hundred times larger, between 150 and 300 million people.[16]

Why did we start farming?

I do not want to imply that the origins of crop and animal domestication can be explained simply and completely as an induced innovation, i.e. an inevitable and gradual response to rising population levels, and that the initial successes of this process enabled and directly reinforced its later expansion and intensification. This book favors facts over convenient narratives.

Few uncertainties in modern science are as deep (and as hard to resolve) as those concerning the origins of the domestication of plants and animals. Evidence has been marshaled and arguments have been offered to explain domestication in starkly different terms, with emphases ranging from purely physical causes to solely behavioral motivations, and even suggesting that the reverse is true—that the plants domesticated us.[17]

A warmer world with higher atmospheric CO_2 concentrations, aka climate change, has been seen as decisive, to the extent that three prominent American anthropologists hypothesized: "Once a more productive subsistence system is possible"—as it was following the last ice age—"it will, over the long run, replace the less-productive subsistence system that preceded it."[18]

Or was it an inevitable reaction to food crises that forced this shift? Perhaps it was a result of the food shortages that arose following the relatively rapid population growth and inadequate food energy provided by foraging? Lewis Binford, an American archeologist, went so far as to put an improbably precise rate as the catalyst for this shift: a population density above 9.098 people per square kilometer triggered the switch from foraging to crop cultivation.[19]

In complete contrast to these physical reasons are the theories that ascribe domestication to the desire for greater socialization and material acquisition (i.e. we wanted more interconnected networks and/or more stuff, so we started farming), and to opportunities for social competition and better organization of defense or offense. A strong argument for the importance of these social components is the fact that the net energy returns of early farming (calculated as the ratio of energy harvested in crops and energy invested in their cultivation) were often lower than the return on foraging activities: in such instances, rewards other than net energy gains were more important.

There is no need for taking sides as far as domestication's origins are concerned. Almost certainly, the process was a combination and interaction of physical and cultural factors, where the old ways of producing food existed alongside the new. But there is no doubt that only the gradual adoption and diffusion of tending the land in one place (so-called sedentary cropping) could support more populous, hierarchical societies with power centered in cities. Chimpanzee-like foraging or combined gathering and hunting endeavors akin to the human activities during the final phase of the last ice age could not support even tens of millions of people globally, never mind that number in a city. The domestication and cultivation of cereals, legumes, oil and fiber crops, and the management of domesticated animals for food and work (pulling plows; carrying heavier loads in transportation) did not eliminate the seasonal and annual variability of food supply, but greatly reduced it through more predictable and far more concentrated production.

Furthermore, this new way of getting food meant surpluses at certain times of the year and the ability to store it for the future. For example, grain in dry regions had naturally low water content and could be kept in suitable containers until the next harvest. This made it even easier for agricultural populations to reach sizes far beyond those attainable by isolated foraging groups. Crop cultivation could support a hundred times as many people per unit of land, even in some of the earliest of such societies. The rate during

Egypt's Old Kingdom (2,700–2,200 BCE) was about 1.3 people per hectare of farmland (that is, 130 people per square kilometer of cultivated land), and it had at least doubled by Roman times.[20]

Eventually, the most intensive versions of traditional cropping in Asia, above all in South China during the late Qing dynasty (1644–1912), relied on irrigation, heavy recycling of organic wastes (mostly animal manures and crop residues as organic fertilizer), the planting of more than one crop a year (using the available land more intensively), and complex crop rotations (ensuring that the soil quality wasn't degraded). These developments supported more than three, and even more than five people per hectare of agricultural land—that is, more than 500 people per square kilometer.[21] By the early 19th century, similar combinations of such improvements meant the English and Dutch could support more than three people per hectare.[22]

More food than ever

Sometime during the first decade of the 19th century, just before widening industrialization and urbanization (and the ensuing rising quality of life that the increased economic output enabled) began to accelerate population growth rates, humanity reached the milestone of 1 billion people. By 2020 that total had grown nearly eightfold. When we look at the global population and the total area of arable land, we see that in 2020 an average hectare of arable land was able to feed five people.[23]

But it is important to note that, while that is the average, global food supply varies widely. It spans large meat and dairy intakes in affluent countries and overwhelmingly vegetarian diets in India and large parts of Africa. The average supply ranges from wasteful surpluses, to widespread malnutrition in the poorest sub-Saharan African nations.[24] In most countries of the European Union and in North America, food energy supply is far above actual needs, and impossible to consume unless entire populations became grossly obese. The results of such surpluses are indeed overeating—with a high proportion of the population being overweight or obese—but also enormous food waste. Meanwhile, many African countries have hardly any supply cushion, and some (Ethiopia being the most depressing example) keep finding themselves repeatedly close to, or beset by, regional or nationwide famine.

What else could we have done?

But was the domestication of crops, dominated by the cultivation of grains and accompanied by the domestication of several animal species, the only possible choice in order to increase both average population densities and the global population total by more than a thousandfold, from a few million to 8 billion—and to do so in the very short evolutionary timescale of just 12 millennia? Or could other food acquisition strategies have resulted in rising populations, a sedentary existence, social stratification, complex societies, and, eventually, a global civilization? Let us look at these alternatives, no matter how unlikely they might appear. But to do that, I must first explain our essential food requirements.

What do we need to eat?

Since the closing decades of the 19th century, we have accumulated substantial evidence concerning human requirements for food energy and for the three macronutrients—carbohydrates, fats, and proteins.[25] Carbohydrates, supplemented by fats, are the main source of food energy, while proteins enable the growth of new and repair of old body tissues (muscles, bones, internal organs, skin), and are metabolized to energy only when the first two macronutrients are in short supply. Moderately active adults should average no more than 2,500 kilocalories (the unit traditionally used by nutritionists and often incorrectly referred to as calories, a unit that is just 1/1,000 of a kcal) or 10.5 megajoules (MJ, in international scientific units) daily—this estimate is on the generous side—and modern dietary recommendations, prepared by international and national committees of experts, specify three optimum intake ranges: 45–65 percent of adult food energy should come from carbohydrates, 20–35 percent from fats, and 10–35 percent from proteins.[26]

The carbohydrates and proteins that we can digest provide about 400 kcal/100 g, and fats more than twice that, at 880 kcal/100 g. Alongside their carbohydrate, fat, and protein contents, freshly foraged plant parts or freshly killed animal and fish meats are mostly water. Fruits provide usually no more than 70 kcal/100 g (and mere traces of proteins or fats; the avocado being the notable fatty exception); tubers (starchy underground storage organs, dominated by potatoes) up to about 115 kcal/100 g; and nuts (because of their high fat content) as much as 650 kcal/100 g. Wild meat from smaller animals contains very little fat—its fresh weight nearly pure protein (about 20 percent) and water, with an energy density of less than 150 kcal/100 g. With this basic information, we can calculate whether there is another practical, long-term alternative to farming that was not, but could have been, pursued. Are there environments in which a large-scale human population could thrive without the need to grow crops?

A world without farming

Eating like a chimpanzee

Could we resort to an intensified version of typical chimpanzee diets and supplement this fruit-dominant diet with other plant tissues, collected insects, and bits of hunted meat? Fruits and vegetables are valuable sources of vitamins and minerals and they provide non-negligible shares of readily digestible carbohydrates in some tropical regions, but an adult American man trying to feed himself like a chimpanzee (with 80 percent of his food energy coming from figs and other fruits and berries) would have to consume 4–5 kilograms of such fresh fruit a day.[27]

That would require, even when living in fig-rich environments, spending hours searching for ripe fruit, climbing trees, and stripping bushes, and eating that many figs every day would add up to 1.5–1.8 tons of fresh fruit a year—and yet he would get no fat and a mere trace of protein. In addition, figs do not grow in higher latitudes, where other wild fruits (from cherries to plums) yield a single and relatively low-yield harvest a year and cannot be stored without processing. In suitable environments, fig-based diets might be conceivable for groups of tens or hundreds of people (much as dates, albeit of a domesticated variety, provide important shares of energy in some African oases), but supplying food energy without any fat or protein for just today's population of the European Union would require more than half a billion tons of figs a year—that is more than 400 times the 2020 global harvest of this fruit.[28] Obviously, any intensified chimpanzee-feeding model is an impossible option for large populations spread from the tropics to the Arctic. Moreover, it is hard to imagine how an existence centered on fig-picking would eventually lead to writing, the Parthenon, and antibiotics.

Eating like a gorilla

Then how about eating just fresh green stems and leaves, plant parts that are vastly more abundant and available for much longer periods in a much wider range of environments than figs or other sweet fruits? Could we have multiplied our numbers by a mass-scale replication of the diet that supports the mountain gorillas inhabiting the slopes of Congo's Virunga volcanoes? The most famous practitioners of eating abundant greenery, these gorillas' total food intake is made up of about 68 percent herb and shrub leaves, and herbaceous stems account for an additional 25 percent.[29] Could we just plump ourselves amid similarly green patches of tall stems and large leaves and start munching?

Even in places where we could do it year-round—and those would not include large regions of the planet, ranging from semi-arid grasslands to boreal forests covered by deep snow for months—adults would have to spend most of the day biting, chewing, and swallowing in order to consume some 10 kilograms of plant tissues a day just to cover their food energy needs. That is if they could metabolize such amounts of plant fiber. But to do that, humans would have to be able to digest it as well as gorillas do, a capability *Homo sapiens* has never had. Our colon (where the digestion of fibrous plant mass takes place) is only about a fifth of the total volume of our digestive tract (it is about half in primates), and it is much shorter than even the colon of chimpanzees, who have more diverse and more digestible diets than gorillas.[30]

Gorillas are hindgut digesters, able to process large volumes of fibrous plant matter, mainly grasses and tree leaves. As is the case with all other hindgut digesters, including horses, rhinos, rabbits, and koalas, their long colons harbor large amounts of fermentative microorganisms (bacteria and anaerobic fungi), facilitating the digestion of a fibrous diet and extraction of energy in the form of short-chain fatty acids. In contrast, our digestive system could extract no more than about 10 percent of the energy present in that fibrous foliage (and most likely less than half of that share), and even after eating 10 kilograms of such greenery a day, we would still be starving. And even if we had more gorilla-like colons, this feeding strategy would depend on the continuous production of new juicy plant mass and be limited to humid and frost-free climates. Again, it is not an option that would lead to the widespread diffusion of settlements, high population densities, and the emergence of cities and a global civilization.

Eating like an Ice Age hunter

What if we took the very opposite route, by moving to a higher trophic level (position on a food web) and resorting to expanded meat-eating? The best (although still limited) way of going down the meat road became impossible once the Ice Age's woolly mammoths and other fatty megaherbivores—including woolly rhinoceros, straight-tusked elephants, aurochs, steppe bison, and giant deer, which could provide large groups of hunters with highly nutritious diets—became extinct. The food energy density of the largest megaherbivore carcasses was 190–240 kcal/100 g but only 120–140 kcal/100 g for smaller herbivores such as deer or antelope. The woolly mammoth and American mastodon weighed 4–9 tons, and the largest imperial mammoths more than 10 tons.

Assuming an average daily per capita need of about 2,200 kilocalories, a 7-ton mammoth (with protein 20 percent and fat 15 percent of its edible carcass, the weight of which would have been about 60 percent of its live weight) could provide a group of 50 people with enough nutrition for nearly 80 days. That is more than two and a half months, and for much of the year there was no problem storing the meat and fat in the low temperatures prevailing at that time. That option is irrevocably gone, however: the last mammoths, trapped by rising sea levels, survived on Wrangel Island (in the Arctic Ocean) and, fatally inbred, died out about 4,000 years ago—that is, half a millennium after the construction of the pyramids at Giza.[31] As we will come on to, the superabundance of smaller grazers could not achieve the same result, which means that the only possibility to match or even surpass those concentrated nutrient levels would be the frequent killing of large marine mammals.

Vilhjalmur Stefansson, the Icelandic Arctic explorer and ethnographer who in the early 1900s lived among the northern hunters in Alaska and Canada, described the difference clearly.[32] Those groups that depended on what he called "blubber animals" (seals, walruses, whales) were the most fortunate, because they never suffered from the lack of fat. In contrast, many forest hunters who failed to kill beaver or moose, and were reduced to eating rabbits, developed the craving for fat known as rabbit starvation, which was eventually marked by diarrhea, headache, vague discomfort, and overall body weakness. Their bellies might have been swollen with rabbit, but they felt hungry.

Large animals hunted by the Inuit or Chukchi (seals, walruses, belugas, narwhals, small whales) are the best examples of the blubber strategy. For example, killing a single 1.3-ton narwhal (25 percent fat, 18 percent protein) would feed a group of 50 people for a month, even with the higher energy needs of Arctic dwellers.[33] We see proportionately similar outcomes with several species of seal, walruses, and beluga whales. But limited numbers of these large marine mammals, difficulties in even locating and killing them, and their restricted distribution to the polar regions would preclude any widespread adoption of the blubber-hunting strategy. As a biologist studying marine mammals noted, "narwhals are hopelessly hard to see, never come when you want them to, swimming far offshore and underwater the whole time . . . Whole field seasons go by and you don't even see a narwhal."[34]

In contrast, killing the largest remaining herbivores such as bison, wild horses, and caribou could provide an excellent source of protein (22–24 percent of their live weight) but very little fat (just 2 percent of live weight in bison) and no carbohydrates. Even the large bison herds that survived in North America into the 19th century were too small to feed tens of millions of people on a sustainable basis. Let us assume that the hunting of the pre-1800 continental herd of 60 million bison—with a (liberal) average live weight of 400 kilograms, boneless edible weight at 35 percent of live weight, and annual killing rate of no more than 15 percent (to maintain the herd)—would yield 1.26 million tons of meat. At about 140 kcal/100 g, that could cover (assuming, improbably, that all protein would be metabolized to carbohydrate) the annual food energy needs of about 2.1 million people, or almost exactly the US population in 1770. Except that most of those people lived east of the Appalachian Mountains, while most of the bison roamed west of the Mississippi. Furthermore, by 1850 the US population was more than ten times as large as in 1770.[35]

And the prospects of feeding at a higher trophic level would be far worse even with a seemingly unlimited supply of deer—the condition that now appears to prevail in parts of North America, where deer numbers have grown by two orders of magnitude during the past eight decades, from an estimated 300,000 animals in 1930 to more than 30 million today.[36] During winter, a group of 50 foragers would need to kill 15–20 deer every week just to get enough food energy. Even a mass slaughter of all of America's 30 million deer would provide food energy sufficient for only about 2.3 million people—which is less than 1 percent of the US population—for a year. (Assuming an average yield of 50 kilograms of meat with an energy content of 120 kcal/100 g—and, again, that all protein would be metabolized to carbohydrate.)

But hardly any fat (deer meat typically has no more than 2 percent fat) and no carbohydrate would make deer-eating a starvation diet. And hunting smaller animals (caught in snares or shot) would be a complete non-starter: a group of 50 people would need to kill nearly 600 rabbits or hares a week (assuming an average meat yield of 1 kilogram per animal, with an energy density of about 120 kcal/100 g), and, once again, in the near-absence of fats and complete absence of carbohydrates in the essentially pure protein diet, this would still leave everybody seriously malnourished. Obviously, going after more abundant but smaller rodents (rats, squirrels, mice) would make—all taste preference aside—the hunting/feeding algebra even less attractive.

Eating like cattle and termites

If we are looking for a metabolic adaptation for organisms of our size and mobility that would allow them to multiply their numbers and reach high populations and high densities, all without resorting to the domestication of crops and animals, there is only one theoretical option left. We would need to do even better than the hindgut-digesting gorillas and be able to metabolize abundant plant mass like ruminants (cattle, sheep, and goats) and termites do. If we are trying to increase the size of the human population, the ability to digest the biosphere's most abundant organic compounds would provide an enormous advantage. These compounds are cellulose, hemicellulose, and lignin, whose large molecules combine to form nature's most common complex compound, making up the cell walls of all plants.[37]

In 2022, the global human population came very close to 8 billion and (considering its age composition and weight distribution) average body mass was about 50 kilograms—that is a total of 400 million tons.[38] To make this total comparable with masses of other organisms, we can express it either in terms of dry weight (human bodies are about 60 percent water, hence about 160 million tons) or carbon (this key element of life accounts for 45 percent of dry mass, hence about 70 million tons). Remarkably, this total is now appreciably larger than for all vertebrates remaining in the wild and only one vertebrate has a collectively larger biomass: cattle.[39]

Cattle (as well as other ruminants, including water buffalo, sheep, goats, deer, camels, and giraffes) can feed on this abundant plant matter because bacteria in their rumen (a part of their four-chambered stomach) produce the requisite enzymes that break down chewed (and rechewed) feed.[40] The domestication of cattle—principally for milk, fieldwork, and transportation (for millennia, oxen, not horses, filled those roles in most societies), not for meat—further increased this advantage, and it eventually made the species the world's single largest repository of animal biomass. And I must stress that having more living mass in cattle than in people is nothing new; it is not a product of the large-scale expansion of modern intensive feeding to produce beef that relies on mixtures of high carbohydrate and high protein feeds (corn and soybeans) consumed by large numbers of animals confined in feedlots. Nearly three-quarters of the world's bovines (living in Asia, Africa, and Latin America) do not receive such concentrated feeds, and even in 1900, long before any centralized animal feeding operations, cattle biomass was larger than the weight of all humans. With about 450 million cattle, it adds up to an equivalent of almost 25 million tons of carbon—about 75 percent larger than the mass of humanity (1.65 billion people) in 1900.

While soil-feeding termites metabolize organic matter present in soil (with no visible plant remains), wood-feeding termites (much like the ruminants) can consume both live and dead wood thanks to the presence of microbes (archaea, bacteria, protists) in their hindgut, which provide the requisite enzymes.[41] We cannot say with acceptable certainty what the global biomass of wood-eating termites is, but a few indicators show how abundant they can be. In the tropics, termites may account for more than 10 percent of the total animal biomass in that environment (and that is in landscapes inhabited by elephants, rhinos, and wildebeest); they may add up to as much as 95 percent of all soil insect biomass; and their densities may reach as many as 1,000 individuals for every square meter of land, as they decompose more organic matter every year than do all large mammalian herbivores.[42]

With maximum densities of about 11 grams per square meter, the biomass of termites in some tropical regions would be 110 kilograms per hectare, almost identical to the average mass of Americans per hectare of cropland in 2020 (330 million people, 55 kilograms per person, 160.5 million hectares). This is not to imply that such termite densities could be supported over a territory comparable to the total area of the United States, merely to indicate the impressive possibility of aggregate biomass resulting from being able to digest the lignin and cellulose making up dry plant matter. But even if hominins had, somehow, acquired termite-like digestion, the global population of such organisms would be limited in numbers and incomparable in terms of socialization, behavior, and accomplishments. Human-size organisms that ate like this could not be like us in terms of behavior or sapience; there would be important downsides. Cattle spend 10–12 hours a day grazing, and humans feeding predominantly on lignocellulosic matter would have to devote considerably more time to feeding than chimpanzees (if not, given the disparity of sizes, as long as bovines or termites).

Even if we had complex stomachs full of cellulolytic enzymes, if we survived on new or decayed wood by foraging we would obviously have to seek out areas rich in these compounds, which are primarily forests and tall grasslands. These adaptations could produce large populations of organisms—larger than chimpanzees—but their individuals would be unlikely to possess extraordinarily large brains like ours; brains that could replicate our current way of living by introducing the widespread mechanical harvesting of wood and grass and its transportation to cities. This means that it would be impossible for these theoretical wood- and grass-eating organisms to urbanize, and although such creatures could attain relatively high population densities, their lifestyle would be akin to that of cattle herds rather than sapient civilization-builders.

Eating like humans

None of the possible intensified foraging strategies, from figs and leaves to blubber and herbivores, would provide sufficient nutrition to grow the human population or allow us to expand into new habitats. Likewise, consuming readily available lignocellulosic compounds was never really an option: the evolution of our digestive system means we lack suitable enzymes. Thus the only mass-scale way of sourcing food that could have led to the larger populations capable of new technological and intellectual advances and our expansion into new areas was . . . farming. Humans' only option—then and now—was to turn to the domestication of crops and animals. Of course, some societies domesticated animals without also domesticating plants, and they succeeded as migratory pastoral societies. But although this strategy provided a nutritious, predictable, and manageable food supply, it could not lead either to high population densities or to substantial permanent settlements—and in the face of other groups that did settle, and which reaped the rewards of doing so (in terms of population size and technical and social innovation), there was huge pressure for migratory groups to follow suit. For one, it's difficult to migrate when settlers claim the land as their own.

It is not about eating as much energy as possible

As we have seen, high population densities and substantial permanent settlements were achieved only by the domestication of crops and animals, with larger, stronger animals used for both work and food. This chapter has shown that energy fuels the growth of civilization—energy fuels the growth of *everything*—and so you might think that human staples—foods that dominate human diets and that are eaten every day, or at least several times a week, for their nutritional composition and digestibility—were selected based on how energy-rich they were. By that measure, sugarcane would be best: this crop beats all other domesticated plants in terms of energy (carbohydrate) yield. And yet no society is built on this as a staple food. Why?

Even traditional, unimproved varieties of sugarcane could produce a massive 40 tons of cut cane stalks per hectare, and that would be enough to yield at least 4 tons of sugar, containing energy sufficient to feed about 20 people for the entire year.[43] That would be a stunning, roughly 200-fold, gain on even the most productive foraging methods. But this option is as theoretical as eating only rabbits, except that the sugarcane route would supply absolutely no protein or fat and little of the essential mineral nutrients. All we would get is a rapidly digestible carbohydrate: sucrose.

Sugarcane was among the relatively early domesticates (in New Guinea some 10,000 years ago), but it became an important global crop only in the early modern era (1500–1800), when trade in sugar—grown mostly on the Caribbean islands and in Latin America—emerged as an important component of European intercontinental commerce, colonization, and the slave trade.[44] After centuries of expansion, sugarcane is now cultivated worldwide on about 27 million hectares (an area about twice as large as Greece), but in 2020 cereal grains—such as corn, wheat, and barley—were harvested on around 740 million hectares (an area nearly 30 times larger), and their global production of about 3 billion tons compared to less than 0.2 billion tons of sugar.[45]

Feeding the masses: our only option

The emergence of the first complex, settled societies; the slow expansion of the global population; the eventual transition to industrial (and now post–industrial) economies; the survival of today's 8 billion people—all of this has depended and continues to rely primarily on harvests of domesticated cereal and leguminous grains. We eat these crops both directly and indirectly: a large portion of the corn, wheat, barley, and soybeans that we grow are used to feed animals to produce meat, milk, and eggs. Arguments about the initial impulse to adopt farming as a survival strategy may never be settled. Was the gradual shift from foraging to agriculture driven by slowly rising population numbers that could not be sustained by hunting and gathering? Was it caused by climate change that made crop cultivation easier; or was it motivated by the desire for material enrichment and private possessions that could be realized only in sedentary societies; or by a drive for dominance and social stratification? Eventually, whatever fueled the rise of farming became irrelevant: once domesticated cereal and leguminous grains were the dominant source of food energy able to support an expanding population and more complex social organization, there was no going back.

No other option offered greater possibilities for such widespread usage—from rice in the Asian tropics and rye in sub-Arctic Scandinavia, to soybeans in insular Japan and corn and beans across most of the Americas—and for securing a more predictable, more storable, and more nutritious way to feed unprecedented population numbers, than grain-based agriculture. In some regions this was supported by tuber crops and oil crops, and vegetables and fruits provided fat, vitamins, and minerals, but it is indisputable that grain-based agriculture fueled the growth of settlements, the emergence and expansion of cities, the development of writing and the arts, advances in exploration, and technical inventions. Many subsequent innovations helped to create today's global civilization—but its primary energetic foundation rests, undoubtedly, on edible grains. Their cultivation was—and remains—our only option.

2. Why Do We Eat Lots of Some Plants and Not Others?

Variety is said to be the spice of life, but where mass scales are concerned, the bulk of human needs are usually catered for by limited sources. For example, three companies (Apple, Samsung, Xiaomi) have two-thirds of the global smartphone market, and four companies (CFM International, Pratt & Whitney, General Electric Aviation, and Rolls-Royce) deliver more than 80 percent of all commercial jet engines. Domesticated plants are no exception to this common rule. Botanists have classified nearly 400,000 species of vascular plants, some 12,000 of them grasses producing small, nutritious seeds; but only a tiny share of all plants have been domesticated. Just 20 species account for 75 percent of annually harvested crops, and today, just two domesticated grasses—rice and wheat—provide 35 percent of global food energy.[1]

Domestication is a deliberate selection and gradual modification of wild species in order to produce plants or animals that are better suited for human needs. This process has been studied (starting with Charles Darwin's description of changes affecting cultivated plants) using a combination of scientific methods, with modern genetics and genomics proving particularly revealing.[2] The eventual outcome of the domestication process is modified plants and animals whose survival and flourishing depend on human care.

Nearly all major crops were domesticated between 13,000 and 5,000 years before the present, in at least seven different regions, starting with wheat, sorghum, millet, rice, potatoes, chickpeas, and peanuts (more than 9,000 years ago), followed by barley, corn, beans, cassava, and sugarcane (before 6,000 years ago), and soybeans, rapeseed, cowpeas, and quinoa (more than 3,000 years before the present).[3]

Archeologists, geneticists, and plant scientists who have studied domestication have identified several notable themes, and the older of the domesticated plants have a larger number of these traits than younger varieties. This is because some of them could have been acquired in just a matter of decades, whereas others have taken a millennium to become established.[4] In wheat, rice, barley, and soybeans, the most important change has been grain retention (so-called non-shattering: when mature grains do not fall out), and wheat, one of the world's two main cereal staples, has significantly larger grains, while domesticated corn and sunflowers have a reduced number of leaves. But the most common traits of domestication are changes of taste (typically to become less bitter, sweeter) and increased sizes of edible parts (larger seeds, fruits, or tubers).

Domesticated fruits offer the most memorable contrasts with their progenitors, be it the wild Chinese gooseberry (small, leathery, and acidic compared to cultivated larger, soft-skinned, and sweetish kiwi fruit) or any stone fruit species: apricots, cherries, peaches, and plums are all much smaller, harder, and less sweet in the wild. Wild bananas are full of small hard seeds, but in cultivated varieties you can hardly see them, as dark specks in the fruit's center. Meanwhile seedless oranges (navels, Valencias, satsumas) have a large share of the global citrus market. The most common consequence of all of these changes is reduced genetic variation.

Domesticated plants include very few cone-bearing (coniferous) trees. The highly resinous bark of conifers and their prickly needles are not fit to eat (although you can make needle tea), and the naked seeds lying exposed on the upper surfaces of female cones are both easily dislodged and difficult to harvest. The price of *pinoli*—pine seeds essential for making true *pesto Genovese*—in a grocery store is testament to the effort involved in collecting them.

About 2,000 species (half a percent of all vascular plants) have been either fully or partially domesticated, and only 200 of those have achieved significant regional or global distribution.[5] For example, how many of these do you know, and how many colors and

Why Do We Eat Lots of Some Plants and Not Others?

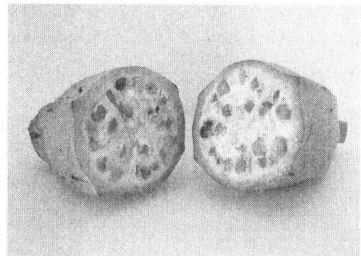 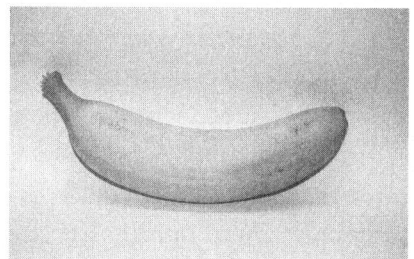

Wild and domesticated: bananas full of hard seed vs. a seedless fruit.

tastes can you ascribe to ackee, biribá, dika, huauzontle, jujube, noni, or pepino?[6] Then there are crops grown for fibers (cotton, flax, hemp, jute, sisal); intoxicants (coca, khat, marijuana, tobacco); and two large groups, herbs (from basil to thyme) and spices (from allspice to turmeric) cultivated for flavor, smell, and coloring, not for nutrition.

Staples: feeding the masses

Some 10,000 years ago, without the benefits of modern chemical analyses or any data comparing the attributes of various species, the selection of staple foods was made without awareness of the macronutrients making up these plants. Rather, domestication was the result of trials, errors, and almost certainly some serendipities.

Suitable seeds or tubers would have been selected because of their unusual size or agreeable taste. Close observations of the germination and ripening of wild plants would have determined the best times for planting and harvesting. Different cooking methods had to be tried to produce more palatable and digestible meals, especially for foods where poisonous compounds needed to be removed (most notably, the soaking and cooking of cassava to remove cyanide). Staples eventually emerged owing to their combination of good yield, widely acceptable taste, good digestibility, and ability to last in storage.

Complex, modern scientific inquiries could not have resulted in better choices than those emerging from this long, experiential selection, which extended across hundreds to thousands of years. It made cereal grains, supplemented by leguminous grains (pulses) and oil seeds, the foundation of human food supply. Millennia later, this food provision pattern remains the basis for feeding the modern world, with some notable modifications that I will explain later in this book. As mentioned earlier in this chapter, fewer than 20 species account for more than 75 percent of annually harvested crops, with the cereal harvest, now approaching 3 billion tons a year, dominant.[7]

But the amount harvested is different to the amount that can be eaten. The largest subtraction is due to significant shares of Brazilian sugarcane and American corn production that are now diverted into making ethanol, used as automotive fuel.[8] The other two large subtractions are due to bagasse (the fibrous residue that remains after crushing cane stalks to yield sugar), which is mostly used as fuel, and so-called oil cakes that remain after pressing oilseeds, and which are usually fed to animals.[9] After these adjustments, cereal and leguminous food grains account for nearly half of the world's edible harvests, and the addition of tuberous staples (white and sweet potatoes, yams, and cassava) and plant oils (sunflower, rapeseed, olive, soybean) raises that share to nearly two-thirds.

With our modern predilection for choice and variety, this highly restricted range of staple crops may seem to be an undesirable and regrettable narrowing of historically richer and more varied diets.

Why Do We Eat Lots of Some Plants and Not Others?

Why did so few species come to account for such a large share of edible harvests? Why are we not habitually eating parts of more than a hundred different plants, as do some chimpanzees or as did many foraging groups, be it in semi-arid environments or in tropical rainforests? Because we needed energy-dense foods, complete with all three macronutrients, that could feed a growing number of people; wheat and rice and corn can do that. This imperative applied in the past, when Neolithic populations evolved from small foraging groups to dwellers of the first clay cities, and it applies today when feeding the largely urban global population of more than 8 billion. But to support those growing numbers in the past as much as to secure enough food for today's record numbers, a food being edible and digestible is not enough to make it a staple—and nor is its rapid growth or easy harvesting. Some 20,000 plants are edible, strawberries and mangoes are not difficult to digest, and spinach leaves are ready for harvesting in less than eight weeks: none of that is decisive.

No, it's a high entry fee to staple food status, and they must meet all the criteria: they must ripen relatively fast; their yield must be fairly high; it should be possible to store them for extended periods; and their digestibility and palatability must be combined with their critical attribute—their ability to supply relatively large shares of necessary nutrients. Domesticated grain staples were able to satisfy all these requirements—albeit not to the same degree in all cases. In Southwest Asia, the earliest group of these domesticates—so-called founder crops—included emmer and einkorn wheat, barley, lentils, peas, chickpeas, faba beans, and flax.[10]

Emmer wheat (*Triticum dicoccon*), with strong husks enclosing its grains, was eventually displaced in most regions by hull-less varieties that are easier to thresh, and it is now a minor crop mostly grown in India and Ethiopia. In the West, the most likely place to come across it is Italy, where it is known as farro (or *farro medio*).[11] Einkorn wheat (*Triticum monococcum*, or *farro piccolo* in Italian) also has tight hulls, and its most famous consumer is probably the man whose more than 5,000-year-old mummified body was discovered in 1991 in ice high in the Ötztal Alps. Analysis of his stomach contents indicated that his last meal was meat and bread baked from einkorn.[12] Faba beans (*Vicia faba*) were cultivated in Galilee as early as 10,200 years before the present day, and ful medames (a stew of faba beans) remains a staple of today's Egyptian cooking.[13] Soybeans, now the world's leading leguminous crop, were a relatively late addition, with a domestication date between 4,000 to 7,000 years ago, in East Asia.[14]

Domesticated corn originated in Mexico's tall and leafy wild teosinte plants about 9,000 years ago, before spreading into South America and across the Rio Grande del Norte.[15] The earliest evidence of today's dominant bread wheat (*Triticum aestivum*)—whose genome comes from three wild ancestors—dates to about 6,400 BCE in southern Turkey, but the first documented wheat with gluten content high enough to make yeast bread is much younger, dating only to about 1350 BCE in Macedonia.[16] Gluten is a plant protein whose elasticity makes leavened bread possible: no-gluten or low-gluten flours are good enough only for flatbreads (leavened gluten-free bread uses psyllium husk as a gluten substitute).[17]

Ancient Egypt's harvests, irrigated by the seasonal flooding of the Nile, rested primarily on emmer wheat and lentils; Roman food supply was based on wheat, barley, oats, rye, and several species of legumes (peas, lentils, beans, chickpeas); the original population of the Americas had a combination of corn and beans; sub-Saharan staples were millets, rice, beans, cowpeas, and chickpeas.[18]

But the most explicit ancient identification of grains as the foundation of nutrition based on domestication is offered by China's mythical Shennong (Divine Farmer) emperor, who is credited with the introduction of *wugu*, or five grains (wheat, rice, soybeans, and two kinds of millet—broomcorn and foxtail), to the inhabitants of North China, thus creating the Chinese civilization.[19] Ancient Chinese historiography places his reign at 2700 BCE, a couple of centuries before the construction of Egypt's great pyramids, but archeological findings make it clear that by that time all of these crops (except soybeans) had been cultivated in China millennia.

Why Do We Eat Lots of Some Plants and Not Others?

Egyptian field work: plowing with oxen, broadcasting seed, harvesting with sickles.

Glorious grains: why they are best

How was this combination of domesticated cereal and leguminous grains able to meet typical food energy requirements while providing desirable amounts of the three macronutrients in readily digestible seeds? How did this convenient natural package obviate the need to eat large amounts of low energy density, and perishable, leafy and fruity plant mass? In order to answer these questions, we need to understand human nutritional requirements and the nutrient composition of these foods.

Food energy requirements are closely related to age, body mass, and overall level of physical activity. Healthy growth in childhood and puberty and the subsequent maintenance of body weight and of essential functions (thermoregulation, metabolism, tissue repair) require that macronutrient needs must be accompanied by an adequate intake of micronutrients (vitamins and minerals). As the qualifiers imply, those needs can be satisfied with small daily intakes, mostly on the order of milligrams or micrograms. A normal diet, consisting of a variety of foodstuffs, is usually able to meet these requirements (and so, on the whole, supplementary vitamins are unnecessary).

As already noted in the first chapter, requirements for the three macronutrients, established by generations of nutritional studies, have well-defined ranges and they also vary with age; obviously, growth or pregnancy will require larger intakes per unit of body mass. The digestible energy of cereals comes mostly from carbohydrates, hard (spring) wheat has a relatively high protein content, and both staples also have a small amount of fats. This combination gives grains relatively high energy densities: wheat (with about 350 kcal/100 g) is about 18 times as energy-dense as an average vegetable (both tomatoes and cabbages contain less than 20 kcal/100 g) and seven times that of common fruit (apples and oranges contain less than 50 kcal/100 g).[20]

To perform some explanatory calculations, let us assume a daily energy need averaging 2,200 kcal for the entire population and 2,500 kcal for an adult working male (males tend to be larger and have relatively more muscle mass, which increases their calorie requirements).[21] Consequently, even if we were able to thrive solely on vegetables and fruits, an average person would have to eat (depending on the species) 5–8 kilograms of vegetables every day, and adult males would have to eat about 18 kilograms of lettuce, almost 10 kilograms of cauliflower, or nearly 4.5 kilograms of apples daily. That would mean munching through some 20 small cauliflowers or eating about 50 apples every day! In contrast, that energy need can be covered by no more than 640 grams of whole grains—that is, about 3.5 measuring cups of rice or just over 5 cups of wheat flour.

But that amount of grain would not deliver enough protein (only about 40 percent of what is needed) and less than half of the needed fats.[22] Recall that modern dietary recommendations specify that 45–65 percent of adult food energy should come from carbohydrates, 20–35 percent from fats, and 10–35 percent from proteins. Eating only wheat or rice could not come close to the two latter needs. This is where leguminous grains made the difference: the universal combination of cereals and legumes has been by far the most remarkable feature of crop domestication, shared by all major ancient civilizations in Europe, most of Asia, and large parts of Africa and the Americas, and preserved throughout the subsequent millennia. But it was only in the 19th century that we understood its biochemical foundation.

A winning combination

Dietary protein, made up of 20 amino acids, is needed for the growth and repair of all body tissues as well as to make up nitrogen losses in feces and urine.[23] We need relatively more of this macronutrient during infancy and later rapid-growth periods. Nine amino acids are considered essential as we cannot synthesize them (i.e. make them ourselves within our bodies) and must consume them in the correct proportions in food. Not surprisingly, all essential amino acids are present in human (as well as cow) milk; indeed, all foods of animal origin (eggs, meat, fish) have more than adequate amounts of all amino acids. They are also easily digestible and allergy to milk proteins is uncommon (no more than 3 percent of the world's population).

However, the most common dietary incompatibility is due to lactose intolerance, the lack of lactase, the enzyme that digests lactose (a sugar). About two-thirds of humanity are affected to differing degrees, ranging from a mild discomfort after consuming a glass of milk to diarrhea and bloating. But, as shown by Japanese or South Korean experience, moderate consumption of milk and dairy products poses no discomfort even among populations with widespread lactase deficiency: those populations now consume every year about 30 liters of milk per capita, compared to just over 50 liters in the EU.

In contrast to animal proteins, all plant proteins are deficient in one or more essential amino acids: cereal proteins are short of lysine, while leguminous proteins are relatively deficient in methionine and cystine, and some are more difficult to digest. And so it isn't precise enough to refer only to a person's overall protein intake; it is important to consider the least abundant amino acid consumed, and it is especially critical to do this in children. We measure the quality of dietary proteins via a digestible indispensable amino acid score: the scores of animal foods are more than 1.0 (1.16 cow milk, 1.14 pork, 1.13 egg, 1.08 chicken breast); those of plant foods are as high as 0.89 (soya flour) and 0.83 (chickpeas), and as low as 0.29 (sorghum).[24]

As a result, mixed diets with significant shares of animal foods always provide more than the needed amount of essential amino acids, but so long as vegetarian diets combine plants like rice and soybeans, for example, they will also score relatively high. The lowest rating (due to lysine deficiency) would come from eating just wheat or only cassava; such a diet could provide plenty of carbohydrates but an inadequate amino acid supply.

This nutritionally optimal mix of cereals and legumes was adopted independently in different parts of the world; in each place it involved different combinations of species but with the same desirable outcome: it increased the amount of quality protein that was consumed.

Without knowing anything about the chemical composition of food and the relative abundance of macronutrients, early agricultural societies remedied the macronutrient imbalance of cereal grains by domesticating legumes relatively early—all legumes are rich in protein, and some are also rich in edible oils and therefore fats. All legumes have a higher protein content than even the most proteinaceous cereal (durum wheat for bread and pasta has a maximum of 14 percent), and in the case of soybeans nearly five times as high as the protein share in rice (typically just 7 percent).

In China, a diet largely comprising whole cooked grains and *doufu* (or tofu—ground and coagulated soybean curd) gave the people access to this winning mix. The combination of rice and wheat (and in the earliest domestication period, also sorghum) with soybeans contains up to 34 percent of digestible energy in protein.[25] In India, dal chawal (lentil rice) is a common combination, with dal containing about 25 percent protein. In Europe, wheat, rye, barley, and oats were combined with peas and beans (protein content of 21–24 percent). In sub-Saharan Africa it was a combination of sorghum and rice with beans and peas: West African favorites include cowpeas (*Vigna unguiculata*, 24 percent protein) and Bambara beans (*Vigna subterranea*, 25 percent protein). And in the Americas, it was not only corn and beans (such as Mexican frijoles con elote), but also peanuts, originally a South American domesticate that was grown as far north as Mexico by the time of the Spanish conquest.[26]

In traditional agricultural societies, meeting carbohydrate and protein needs with a combination of cereal and leguminous grains was made easier by the fact that less protein and less fat was consumed than is nowadays recommended. Much as the Inuit could survive winters with hardly any carbohydrates (in summer they collected berries, grasses and stems, roots, and seaweeds), protein supply was often near the lower range of today's recommended optima, and fat could be even scarcer. The best, and by far the most accurate, documentation of this comes from an incomparably detailed farming and nutrition survey performed under the leadership of John Lossing Buck, an American agricultural economist, of China in the late 1920s and the early 1930s.[27]

At the time the country's agriculture and diets remained very close to the situation that had prevailed under the Qing, the last imperial dynasty (1644–1911), and extensive surveys in 136 localities in principal cropping regions provided both representative and accurate results. The nationwide average showed that adult males derived 83 percent of all food energy from cereal grains, nearly 7 percent from legumes, and 2 percent from plant oils, making seeds the source of 92 percent of all food energy; other plant foods (in order of importance: potatoes, vegetables, cane sugar, and fruits) supplied 6 percent, leaving a mere 2 percent for foods of animal origin. How unchanged the cereal share had remained can be illustrated by the fact that in Ming dynasty China (1368–1644) the expected annual per capita supply was 3.6 *shi* of staple cereals (rice and wheat).[28] With 1 *shi* being 60 kilograms, that is 216 kilograms a year or 590 grams a day—and hence about 2,100 kcal/day, or almost 85 percent of a working man's daily energy need, the total virtually identical to Buck's 1930s average share.

Similarly high, or only slightly lower, reliance (75–80 percent) on cereal and leguminous grains was the norm in India and in large parts of early modern (1500–1800) Europe. And even in the rapidly urbanizing and industrializing Britain of the 1860s, carbohydrates supplied nearly 70 percent of all food energy for an average family.[29] During the years of good harvests, the combination of cereals and legumes was able to provide not only enough carbohydrates but also the minimum protein requirements (about 50 grams a day for adults). Although, as I will explain later, given the low intake of meat and dairy products, the protein quality was below optimum.

Cereal and leguminous grains have other advantages besides their relatively high energy density and desirable combination of macronutrients. They provide a good supply of several micronutrients—above all B-group vitamins, copper, iron, magnesium, phosphorus, and zinc—and nutrients lost during milling are now commonly not just replaced but their presence enhanced by compulsory fortification, i.e. enrichment by minerals and vitamins.[30] Ripe seeds have a low moisture content (less than 15 percent) and can be stored in smaller containers or larger granaries for months, in proper conditions even for a few years, and they are also easily transported—traditionally in sacks, and since the 19th century in increasing volumes intercontinentally as loose bulk loads in large ships. These two realities are best illustrated by the shares of annually harvested cereals that are stored and traded: in the early 2020s global grain stocks were equivalent to nearly 30 percent of the world's annual cereal harvest, and nearly 20 percent of the annual grain harvest (and about a quarter of all wheat) was internationally traded.[31]

Cereal processing yields a large variety of foodstuffs. By far the most important are flours (milled from wheat and from rye), rice, and corn. They are used to bake bread (originally the staple prepared food of Europe and parts of Asia, now a leading global baked product), tortillas, pastries, and noodles—from the now globally available Italian pasta to many East Asian wheat and rice (and in Japan also buckwheat) varieties. Grain-based alcoholic beverages go back to ancient Egyptian beers—made, as today, from fermented barley—and include Japanese sake, made from rice. Traditional meat substitutes widely eaten in East Asia (tofu is at least two millennia old and is made from coagulated ground soybeans; seitan is made from wheat gluten) have become newly popular with the fashion for veganism in affluent societies. Grain-based condiments (vinegars, soy sauce) are now globally distributed, and corn (just 3–6 percent fats in the whole grain, concentrated in the embryo) and two leguminous grains (soybeans 18 percent and peanuts 48 percent fats) are sources of cooking oils.

Shortages

These relatively oily seeds are exceptions among staple grains, and in most premodern agricultural societies, whose diets were dominated by carbohydrates, fat was the macronutrient in the relatively shortest supply. Rather than today's recommended minimum of 20 percent of all energy, it provided as little as 10 percent, contributing to undernutrition and stunting. Not surprisingly, all kinds of fats (butter, fatty meat, cooking oil) were cherished by most people in premodern societies, and in some places severe fat shortages were eliminated only recently. During the 1970s the cooking oil ration in most Chinese cities was just 100–200 grams a month; to stir-fry just two dishes for a frugal family meal requires about 50 grams (roughly 4 tablespoons) of oil.[32] And in war-destroyed Japan of the 1950s, fats supplied even less food energy than vegetables. Subsequently, their intake tripled before declining again, as the aging (less active) population requires less food energy.[33]

While the cereal–legume combination was able to cover minimal macronutrient needs in all intensively farmed regions, there have been repeated shortages and even famines arising from wars, epidemics, and weather-induced crop failures. Other factors compounded this.

First came the need to save a portion of the harvest for the next year's seed. This share depended on the yield, and poor grain harvests in early medieval Europe meant as much as 50 percent of the harvest had to be saved as the next year's seed; with high yields, the rates were 25–30 percent.[34] Modern producers do not need to worry; they do not plant seeds reserved from the previous year's harvest but buy new seeds annually from specialized growers.

Next were the losses owing to inadequate grain storage, ranging from spoilage due to high moisture to infestations of insects and rodents. In order to minimize so-called shattering losses in the field (preventing dry grains dropping off), staple grains are harvested when their moisture content is between 22 and 25 percent. Today, farmers across the world can accurately measure this and know exactly when they should pull the trigger on the harvest. Once harvested, water content should be less than 15 percent even when storing grains for just a few months, and below 13 percent when storing for the longer term.[35] Pre-storage drying—in many poorer countries still mostly natural (by spreading grains on sunny surfaces, leading to contamination and insect infestation)—is thus essential to prevent mold, as well as to reduce transportation costs (moving less water weight). Mechanical drying is preferable but relatively expensive, and hence still beyond the reach of small producers.

Grain losses in traditional storage structures (made of mud, wood, and grasses) could amount to half of the initially stored mass, and while the modern storage of properly dried grain in silos and steel bins has reduced these losses to less than 2 percent or even 1 percent, in some countries the aggregate losses across the entire supply chain (from harvesting to retail level) remain high. For example, the totals for rice average about 15 percent in China (estimates ranging from 8 percent to 26 percent) and Thailand, and as much as 25 percent in Nigeria, while the rates of post-harvest losses for wheat in India range mostly from 4 percent to about 12 percent, and are up to about 15 percent in sub-Saharan Africa.[36]

The actual consumption of harvested wheat and rice is further reduced by their milling. Whole grains, of both wheat and rice, can be eaten after cooking, and among poorer Asian populations unmilled brown rice was a common staple (in Japan augmented by whole grains of barley, in China by millet). But a few relatively minor exceptions aside (above all bulgur, and cracked and parboiled wheat grains), wheat has been always milled to remove the outer shell (bran) and the germ (embryo), and then the endosperm (tissue surrounding the embryo) is ground to different levels of fineness. Milled grains are more palatable but low yields precluded their general adoption: brown rice (with only hulls removed) dominated Japanese rural diets until the beginning of the 20th century. While storage losses are highly variable, milling losses can be anticipated. Extraction rates (flour share produced from a unit quantity of whole grain) can range from nearly 100 percent for whole-wheat flour to only about 70 percent for American all-purpose white flour.[37]

Rice processing often results in even greater losses. The endosperm makes up 69 percent of rice grains but only a very small share of harvested rice is milled into flour, used to make noodles and rice paper. The extraction rates of milled white rice are typically 68–72 percent but, depending on the grain variety, yields in small mills are often as low as 50–60 percent.[38] As a result (after subtracting harvesting, set-aside seed, and storage losses), less than half of harvestable grain would be available for human consumption even when eating brown rice or whole-wheat and rye bread, and the share would be less than 40 percent after milling.

Terrific tubers

There is only one carbohydrate staple that could be harvested when needed for immediate consumption or for short-term storage: cassava (*Manihot esculenta*, known also as manioc or yuca), a perennial tropical tuber now grown as an annual. The tuberous root is very starchy (97 percent carbohydrate) with mere traces of protein and fat, but its annual output (recently about 300 million tons a year) ranks it right after rice and corn as a leading staple carbohydrate throughout the tropics, with Nigeria (about a fifth of the world total), Brazil, India, Angola, and Ghana being its largest producers.[39] Cassava is eaten boiled or used to make (by grating, expressing moisture, and drying) coarse flour (farinha, garri), and its main positive attribute is that it does not have a specific harvesting period: harvests can be delayed for months, effectively storing the crop on site, although they eventually become fibrous and woody.

Cassava is ready for harvesting at any time between six months and two years after planting. Harvesting for human consumption usually takes place when the plants are 8–10 months old, although longer growing periods generally produce a higher root and starch yield. Commercial harvesting is done mechanically, but small-scale growers can do it as needed, plant by plant: just cutting off the stem about half a meter above ground, loosening the surrounding soil, pulling the plant up, and cutting off the tubers.

In contrast, potatoes, the world's leading tuberous carbohydrate staple, must be harvested within a narrower time span at maturity and their prolonged storage often suffers higher losses (they dry out, sprout, or spoil through diseases) than those of cereal grains.

Andean populations, the original potato domesticators (starting some 8,000 years ago around Lake Titicaca, between today's Bolivia and Peru), solved the problem with a region-specific (cold, high-altitude climate) solution. Dehydrated chuño was made by spreading the potatoes out on straw beds and freezing them overnight; the subsequent expression of moisture was carried out by stepping on the tubers, and the sequence was repeated until the potatoes were dehydrated, resembling small stones, and could be stored for years before rehydration.[40] Convenient and helpful as this storage method was in the cold Andean climate, potatoes were not the dominant staple of the great Inca empire—or for that matter of any high civilization that left behind remarkable monuments—a combination of corn and beans was.

Societal impacts

Besides providing the largest share of macronutrients for expanding populations, domesticated grains impacted the social, economic, and technical development of complex civilizations. The domestication of crops required unprecedented foresight and management measures. Because staple food tissues (seeds, tubers, nuts, fruits) overwhelmingly take a long time to mature and commonly yield just a single harvest per year, humans had to become long-range planners and managers. This required seeding in time, harvesting at the most appropriate stage of ripeness, producing enough to last until the next harvest, processing the gathered biomass to make it more palatable and storing it with minimized losses, and setting aside enough for the next year's planting.

No staple grains can mature as rapidly as some vegetables (lettuce is ready to harvest in just 30–40 days; cucumbers, beets, broccoli, and zucchini in 40–60 days); staple grains are annual crops, planted and harvested once a year, requiring between 90 days (spring wheat in Canada is planted in May, harvested in August) and 120 days (Japanese rice, depending on the region, is planted in April–May, harvested in August–October) to mature, which means that in climates suitable for their cultivation they are limited to a single crop a year, usually with spring planting and a late summer/early fall harvest in the northern hemisphere.[41] In Europe, North America, and China, winter wheat as well as winter barley, oats, and rye, planted in fall and harvested the next summer, are major exceptions: annual crops whose harvests come during the subsequent calendar year.

Legumes have similar maturing spans: soybeans in North America are planted between May and June and harvested between September and November; for China, the span starts and ends a month sooner. In monsoonal Asia, planting in one year and harvesting during the next is common: India's rabi (winter-sown) crops of wheat, rice, sorghum, and corn are planted between October and December and harvested between March and May. Tubers also have similar growing calendars: both white and sweet potatoes are ready in 90–120 days. Fluctuating labor demands (high for transplanting and weeding, and especially critical for timely harvesting) imposed further production limits that were removed only by the mechanization of field tasks (tractors pulling plows and seeders and fertilizer applicators; combines and other harvesters), which took place gradually—starting in the early 20th century in North America.

Post-harvest measures—saving seed for planting; storing seed until the next harvest; milling wheat, rye, rice, and corn—required advanced planning, and as the population and food demand increased, they led to important innovations and technical advances. Granaries were built by many centralized and well-administered ancient states including ancient Egypt, the Roman empire, and imperial China. During the end of the 18th century in Qing China, annual grain stocks were fluctuating mostly between 2 and 2.5 million tons.[42] That translated to as much as 10–15 kilograms per capita and, undoubtedly, this extensive and well-organized buffer against famines helps to explain why Qing China was such an accomplished economy.

Continued growth

The ancient origins of the need for imported grain are most famously illustrated by large-scale imports of wheat to Rome, the city whose population at the beginning of the Common Era required about 200,000 tons a year, much of it freely distributed and most of it shipped from Egypt and North Africa. This procurement, processing, storage, and distribution required a great deal of planning and coordination.[43] Twenty centuries later we ship wheat in such quantities and rely on such high-capacity vessels that just three or four modern bulk cargo carriers could transport Rome's annual grain import to its destination. And China's tradition of large-scale grain storage reached new highs in 2021, when the country with some 18 percent of the world's population amassed more than half of all global grain stocks, including stockpiles of wheat that could meet Chinese demand for one and a half years.[44]

The need for larger-scale milling of grains (and oil pressing from seeds, initially done manually or using animal-powered millstones) was a major reason behind the development and diffusion of continually improving waterwheels and windmills, and the overseas grain trade stimulated the development of cargo shipping—until the advent of railways, a preferable choice for long-distance imports than any terrestrial transportation. As the population increased, processing and shipping capacities kept pace. The world's largest mills can now process several thousand tons of wheat on a single site every day, and the world's largest grain ships can carry more than 50,000 tons of wheat from Russia (now the world's largest exporter) or soybeans from the US and Brazil.[45]

The downsides

Given the duration, scale, and intensity of this transformation in the world food system, the evolution of a grains-dominated agriculture has had many inevitable negative consequences. The main concerns range from reliance on diversity-reducing monocultures (planting the same crop year after year) to the restricted range of staples in modern diets; and from degradation of soils (soil erosion, soil compaction by heavy machinery, irrigation-induced salinization) to the environmental damage caused by fertilizers (loss of soil organic matter, contamination by heavy metals); and, most recently, agriculture's contribution to global greenhouse gas emissions. Some of these consequences, such as the long-term shift toward cultivating fewer species in larger quantities, have been inevitable given the scale of demand and the universal trend toward urbanization, which requires more intensive cultivation of staples. Others can be significantly remedied by better farming practices or much reduced by modifying our diets, or by eliminating clearly questionable practices. I will have more to say on these matters later.

Farmageddon: our biggest mistake?

For some anthropologists and historians, this has not been enough. Jared Diamond, author of popular global histories, pronounced agriculture, "the worst mistake in the history of the human race . . . Forced to choose between limiting population or trying to increase food production, we chose the latter and ended up with starvation, warfare, and tyranny," and he faults agriculture for bringing us malnutrition, epidemic diseases, deep class divisions, and gender inequality. Humanity became trapped in bottomless agricultural misery—while hunters and gatherers, according to Diamond, "practiced the most successful and longest-lasting life style in human history," lived better lives, and were healthier, taller, and happier than any farmer could be.[46]

Such extreme and sweeping historical misjudgments are based on a few questionable studies of some foraging societies that survived into the 1960s. These were first widely publicized by American anthropologists at the 1966 conference "Man the Hunter," where hunter-gatherers were elevated to the status of "original affluent society."[47] This fashionable infatuation found its new ideal of a so-called noble savage (uncorrupted by civilization, devoid of any vices) in the remaining foragers of the Kalahari and in other surviving small groups, portraying them as supremely satisfied societies living conflictless lives of leisure, providing for themselves with minimal exertion and enjoying health and fulfillment.[48] Subsequent critiques by more discerning anthropologists demonstrated how these conclusions were based on very small groups and time-limited sets of observations, but not only that: questionable definitions of such key terms as "work," "leisure," and "affluence."[49]

Subsequent—and clear-sighted—anthropologists have also documented how the claims of affluence ignored the foraging groups' repeated experience of malnutrition, food insecurity, premature death, short life expectancy, and infanticide, as well as extraordinarily high rates of violence. No matter, the misleading mantra of harmless, sharing, satisfied, and well-fed foragers became widely accepted, which tells us more about both the proselytizers of this wishful nirvana and their receptive audience (their underlying yearning for pristine environments; their disparagement of modern societies) than about the real world of hunters and gatherers.

Even so, uncritically idealizing foraging societies and demeaning the domestication of crops and animals is one thing, but to imply, as Diamond did, that agriculture did not encourage any flowering of art—because "gorillas have had ample free time to build their own Parthenon, had they wanted to"—is quite another.[50] Diamond does not explain how those Parthenon-capable gorillas would quarry stone, move it, shape it with precision, plan astonishingly symmetrical structures, and build them: setting aside their inadequate number of neurons, how would quadrupeds accomplish such tasks? Hyperbolic statements are one thing, Parthenon-building gorillas quite another.

As Jean d'Ormesson, French philosopher and writer, noted, "History cannot advance without people, plenty of people."[51] And having plenty of people is clearly predicated on domesticating crops and animals and on adopting, diffusing, and intensifying agriculture—above all its dominant grain-based variant. Diamond would prefer to have no agriculture, hence no history and just some small, scattered, vulnerable groups of foragers—or, perhaps even better, just some gorillas that chose not to waste their ample free time on anything more complicated than eating leaves and slow procreation.

And then there is James C. Scott, an American political scientist and anthropologist who, in his book *Against the Grain*, holds grain-based cultivation responsible for all the ills of the earliest states. Such states, he claims, were defined by poor nutrition, impoverishing taxation, subjugating population control, vulnerability to disease, and propensity for social collapse—as opposed to the "rather good" life of pre-agricultural "barbarians."[52] Scott also points out that plant and animal domestication preceded true sedentary societies and that the narrative of domestication creating settlement is wrong. Not so simple: archeological findings have documented that proto-domestication and domestication coexisted for millennia with all forms of foraging, that some semipermanent or permanent settlements appeared and grew even before the adoption of regular crop cultivation (Turkish Göbekli Tepe being the most famous example), and that some incipient small-scale crop harvesting was done by foraging societies. Nothing else could be expected in a protracted, multifaceted process unfolding in a wide variety of environments and changing climates, and among so many different tribal societies. Clearly, changing the way the world feeds itself has been a hugely complicated process that cannot be encompassed by a caricatured narrative.

Death by wheat?

Recent decades have seen yet another facet of grain's demonization—above all, the elevation of wheat in general, and highly milled white flour in particular, as the worst imaginable source of modern health problems, from relatively rare celiac disease to the rising incidence of diabetes. In 2015, William Davis, an American cardiologist, brought this to another level, claiming (in a *New York Times*–bestselling book, no less) that wheat has killed more people than all wars combined (!) and that it causes a large array of health problems including not only obesity and diabetes but also heart disease, dementia, schizophrenia, and cancer.[53] Perhaps wheat is also responsible for all atrocities and natural disasters.

But then, how to square this litany of harms with actual wheat consumption and longevity? How is it possible that a lifelong daily intake of this pernicious foodstuff—supposed to cause so many serious life-shortening diseases—is associated with the world's most enviable life expectancies? Those who know something about Italian food appreciate that Italian bakers insist on the finest (highest-extraction) white flour—Farina 00—to make their pasta and pizza, and it is easy to check food supply balances and life expectancy data and see that Italians are consuming annually 55 percent more wheat and highly milled flour than Germans (who like their darker Bauernbrot with a bit of rye), but that their average life expectancy tops the German mean by two years.[54] Are the debilitating health effects of the finest Italian flour counteracted by climate or language or different hand gestures?

Spain provides another convincing proof of how baseless Davis's identification of wheat as a supreme killer is. During the 1960s, in the early stage of Spain's delayed modernization, per capita wheat consumption declined by about a third as Spaniards began to eat a more varied diet, but afterwards it has remained high and little-changed, averaging annually 101 kg/capita in 1970 and 100 kg/capita in 2020—yet during those five decades the country's combined life expectancy increased by about 12 years.[55] That life extension has come close to the record gain in rice-eating Japan (13 years), where wheat supply rose by more than 10 percent during that period—and significantly (about 20 percent) higher than in Canada, where recent wheat consumption has been about 10 percent lower than in Spain. Moreover, besides Japan, Italy and Spain have the highest longevities among the world's populous (>10 million) nations.

And comparisons of life expectancy among Europe's healthiest nations offer yet another, and very convincing, debunking of

Davis's preposterous killer-wheat claim. Except for Germany, all those nations with average longevities above 81 years are heavy and lifelong consumers of wheat eaten daily as bread, pasta, and countless savory and sweet items ranging from crunchy grissini and pretzels to creamy gateaux and tortas. Moreover, three of the continent's five countries with the highest life expectancies (above 83 years)—Switzerland, Italy, and Norway—are exceptionally heavy consumers of wheat-based foods, with Norway and (as already noted) Italy consuming about twice as much per capita as Germany!

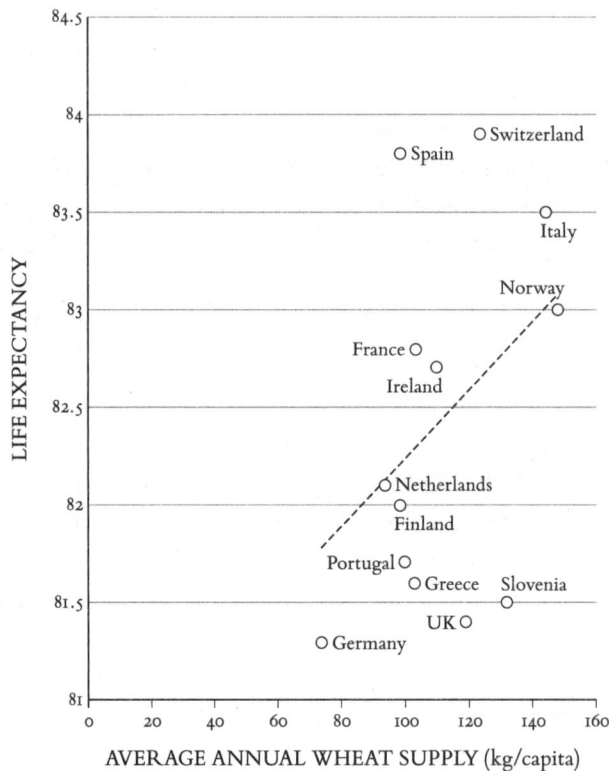

Life expectancies (in years) and average annual per capita wheat supply (all data are for 2022).

Contrasting facts and extreme claims

Setting aside all biases, demonization, and unsupportable claims, the quantitative and qualitative evidence suggests some indisputable facts. Only domesticated crops could provide the energetic and nutritional foundation for the continued (and continuing) growth of human populations and for their economic, social, and cultural development. Only the combination of cereal and leguminous grains supplemented by tubers and oilseeds could provide the predictability and reliability (albeit still with risks attached) of staple food supply, prolonged storability of surpluses, and possibility for long-distance staple food trade. Only adequate food supply (minimizing the recurrence of frequent and prolonged famines) could support (for better or worse) the growth of human technical and artistic achievements. Only the eventual improvements of mass-scale grain-based nutrition could lead to unprecedented gains in longevity and quality of life. To deny any of this is to ignore the most fundamental of all human physical and material realities.

Although on the global scale, the transformation took millennia to accomplish, on the evolutionary scale it progressed rapidly.[56] At least 56 million years elapsed between the time of the first primitive primates (66 million years ago) and the separation of gorilla ancestors (perhaps as early as 10 million years ago) and, soon afterwards, of chimpanzees. The first bipedal primates appeared about 5.8 million years ago, the first member of the genus Homo about 2 million years ago, and *Homo sapiens* has been around for about 300,000 years, while it took only about 7,000 years from the end of the last ice age and the beginnings of crop domestication to the appearance of the first grain-based states. But this is not the story of a straightforward ascent of grain—this evolutionary acceleration had to overcome several fundamental natural limitations, as we will see in the following chapters.

3. The Limit of What We Can Grow

All food begins with photosynthesis, the biosphere's most important reaction. Photosynthesis produces new plant mass (phytomass) that can be consumed as is (lettuce, strawberries), after simple mechanical preparation (the peeling of vegetables or cracking of nuts), or after more elaborate processing (the milling of grains; the fermentation of milk to produce yogurt or of soybeans and wheat to make soy sauce). Plenty of phytomass unfit for eating (leaves, straw, milling residues) makes excellent animal feed in order to produce meat, dairy foods, eggs, and now also farmed fish. And mushrooms are not an exception: of course, they do not photosynthesize, but they derive their nutrition by using their mycelia (thin filaments) to tap tree roots, dead logs, and stumps and debris in the wild, and logs, straw, sawdust, or manure in commercial production. They, like meat or eggs, are indirect outcomes of photosynthesis. Of the mineral nutrients essential for a healthy life, only the minerals (from iron to zinc) are not the product of photosynthesis, but many foods contain amounts of these essential macro- and micronutrients absorbed from soil.

Photosynthesis uses incoming solar radiation to energize the synthesis of new plant mass from carbon dioxide, water, and essential macronutrients (nitrogen, phosphorus, potassium, calcium, magnesium, and sulfur) and micronutrients (ranging from boron and manganese to molybdenum and zinc).[1] Carbon dioxide is abundant and its concentrations have been rising, mainly as a result of fossil fuel combustion, but they remain below levels that would fully optimize photosynthesis: that is why greenhouses have atmospheres that contain twice or even three times the current atmospheric CO_2 level of just above 420 ppm (parts per million; that is, 0.042 percent).[2] But CO_2 does not present the most worrisome limit on photosynthetic production; that comes commonly from shortages of water and nutrients.

As it turns out, it is exceptionally rare that plants have access to these photosynthetic inputs in optimum amounts; and even if they had an adequate supply guaranteed, they still could not produce new phytomass with high efficiency because this life-sustaining reaction, taking place everywhere outside of permanently frozen or extremely arid environments, has a fundamental, inherent drawback: it is surprisingly inefficient. This matters, because the quest for higher efficiencies has been one of the fundamental markers of technical and economic advances. When this effort is applied, overzealously, to labor, it can lead to the unconscionable exploitation of humans' capacity for work. When it aims at improving the performance of everything from individual machines to complex engineering systems, this quest—especially when followed in a persistent, methodical fashion—is most welcome, as it leads to less waste, reduced environmental impacts, higher profits, and more reliable performance.[3]

Throughout history we have been on a quest to raise the efficiencies of energy conversions—we want to maximize useful energy output from a given energy input. This process found new fervor in the Industrial Revolution, with the invention and diffusion of new forms of energy conversion. But as important as these were, none were as important as the concurrent increased efficiencies in how we produced our food.

Energy, engines, and efficiencies

Before 1720, Thomas Newcomen's first coal-burning steam engines (used to produce mechanical energy, mostly for pumping water from deep mines) were so inefficient—converting less than half a percent of coal's chemical energy into the mechanical energy to power a pump—that they could be used only at pitheads, where the fuel was cheap because the machine was located at the source of the fuel, requiring no transportation; plus, if you owned a pit, then you were not short of fuel. During the 1780s, James Watt's improvements (above all, a separate condenser that left the steam cylinder hot) raised the efficiency only to about 2 percent; but by 1900 some massive stationary steam engines had efficiencies surpassing 15 percent, and the best steam locomotive engines were more than 6 percent efficient.[4]

Internal combustion engines, powered by gasoline and diesel, far surpassed those performances. Gasoline engine efficiencies rose from more than 10 percent during the 1890s to more than 20 percent by the 1930s—a time when diesel efficiencies were surpassing 30 percent. Recently the best diesels (large machines that power intercontinental shipping) have reached efficiencies in excess of 50 percent, and the best natural gas–fired combined-cycle gas turbines (which combine gas and steam machines) became the most efficient internal combustion engines, with efficiencies up to 65 percent. The lighter kerosene-fueled gas turbines that propel all modern jetliners have efficiencies of more than 40 percent, and, with proper maintenance, can operate for more than two decades.[5] And while early oil-fired furnaces for household heating converted only half of the fuel's energy content into useful heat, the best natural gas–fired furnaces now have efficiencies of about 97 percent, second only to resistance electric heaters operating at 100 percent efficiency.[6]

Efficiencies of electricity conversion have followed a similar trend. In 1882, Thomas Edison's first light bulbs with carbon filaments converted just 0.2 percent of electricity into visible radiation. By 1900, better filaments raised that only to about 0.5 percent. The introduction of fluorescent lights during the 1930s increased efficiency to about 15 percent, and today's light-emitting diodes convert 80 or even 90 percent of all electricity to visible light.[7]

The closest anthropogenic counterpart to photosynthesis is the conversion of solar radiation to electricity by photovoltaic (PV) cells. The maximum theoretical efficiency of traditional single-junction PV cells (known as the Shockley–Queisser limit) is 33.16 percent.[8] The efficiencies of the commercially available cells of the early 2020s are nearly 20 percent for crystalline and just 6 percent for amorphous silicon.[9]

Our bodies are also energy converters. We take in the chemical energy of food and turn this into heat to maintain constant body temperature and into the mechanical energy of beating hearts and muscular exertion. But our muscles are much less efficient converters than today's best engines or PV cells. Thanks to classic experiments by Francis Benedict and Edward Cathcart in 1913 that measured work during exertion—due to running on a treadmill or pedaling—we have known for more than a century that during aerobic exercise the muscles of well-trained athletes convert chemical food energy to mechanical energy with efficiencies of 16–21 percent.[10]

Photosynthetic fundamentals

How does photosynthesis compare with these achievements? A standard introduction to photosynthesis, as you might encounter in schools around the world, conveys no hint of its inherently low efficiency. Open a biology textbook or search "photosynthesis" on the Web, and you get the process equation written as:

$$6CO_2 + 6H_2O \rightarrow C_6H_{12}O_6 + 6O_2$$

This looks simple, neat, efficient, and thoroughly beneficial: plants take up six molecules of carbon dioxide (CO_2), a trace gas present in the air, and six molecules of water (H_2O), absorbed by their roots and transported to the leaves, and use solar radiation to synthesize a molecule of glucose ($C_6H_{12}O_6$) while releasing six molecules of oxygen (O_2). But this frequently reprinted equation offers only a very simplistic glimpse of a vastly more complex reality. As already noted, photosynthesis is a very intricate sequential process that requires other material inputs besides CO_2 and water, produces a range of organic compounds, and its surprisingly low efficiencies of converting solar radiation into new plant mass are even lower when we express efficiency in terms of edible products. Even after excluding all roots that plants produce for structural and functional reasons, only a portion of the harvested above-ground phytomass of cereal, legume, and oil crops ends up as food. Photosynthetic efficiency—expressed as the percentage of solar radiation received during the growing season that was converted to the chemical energy of a newly grown plant—is merely in the category of Newcomen's or, at best, Watt's 18th-century steam engines, rather than resembling modern turbines, or even trained muscles.

Three pathways

Plants follow three photosynthetic pathways—or, more accurately, one pathway that in some species is preceded by important biochemical and structural modifications.[11] The first is globally dominant and it is found in most staple cereal and leguminous grains, as well as in all tubers, oil crops, vegetables, and fruits. The second is represented among major food and feed crops only by three grain species (corn, sorghum, and millet) and by sugarcane; and the third is limited to succulent plants, which, in food or drink terms, means only several fruits of cacti (above all *Opuntia ficus-indica*, prickly pear), pineapple, and the fleshy leaves of *Agave tequilana*, which are cut and fermented to produce tequila.

The world's most common photosynthetic pathway, deployed by wheat and rice—by far the most important staple grains globally—begins with the synthesis of phosphoglyceric acid (PGA). This reaction is possible thanks to an enzyme—ribulose-1, 5-bisphosphate carboxylase/oxygenase (Rubisco)—that is able to catalyze the reaction of CO_2 with a sugar phosphate, and this ubiquitous function makes it the most abundant protein in the biosphere.[12] Subsequent steps in the photosynthetic cycle produce glucose, but this simple sugar, readily soluble in water, cannot form any durable plant tissues and hence it must be polymerized (ringed glucose molecules are combined into long chains linked by oxygen) to produce cellulose, the biosphere's most common organic compound.

Glucose must also combine with nitrogen, obtained from dissolved nitrates taken up by plant roots and synthesized into amino acids that form proteins—irreplaceable structural and functional components of all living organisms. And many crops need additional energy and material inputs to produce the lipids stored in their seeds. Full comprehension and appreciation of these fundamental reactions requires an understanding of the requisite physics and biochemistry. Fortunately, there is an easy way to calculate the surprisingly low efficiency of crop photosynthesis: anybody able to do basic arithmetic can do so after accessing just three kinds of readily available data on the internet.

The first task is to find the average (or maximum) harvests of a crop in a specific location. For example, Manitoba, where I live, produces some of the world's best spring durum wheat, known for its high protein content and hence perfect for making pasta. Recent typical yield has been about 50 bushels per acre, or about 3,300 kilograms per hectare.[13] (American non-metric measures remain entrenched in the continent's cropping.) The second task is to convert this mass into its energy equivalent (energy contents of all staple cereals cluster around 16 megajoules per kilogram) to get the yield (edible energy output) of some 53 billion joules (53 gigajoules) per hectare. The last task is to find the energy input: the total amount of solar radiation that reached one hectare of that wheat field during the crop's growing season.

For Manitoba wheat that season is short—just 90 frost-free days in the three summer months. Monthly data showing global solar radiation are available from various compendia, but perhaps the most convenient source is the Global Solar Atlas, where you can find detailed monthly (also daily and hourly) irradiation data for any locations with the exceptions of southernmost Argentina and Chile.[14] In southern Manitoba, a 1-hectare wheat field receives about 20 trillion joules (20 terajoules) of solar energy during the growing season. The quotient of output/input (53 gigajoules/20 terajoules) is 0.00265; only about 0.27 percent of solar energy that reached a hectare of that wheatfield during 90 growing days was converted into the chemical energy of harvested grain. That is two orders of magnitude below the conversion efficiency of the best commercially deployed photovoltaic cells (about 20 percent). Why is there such an enormous disparity?

Why is photosynthesis so wasteful?

Theoretically, the efficiency of basic sugar-forming synthesis is about 27 percent, similar to that of working muscles or to the performance of the best photovoltaic cells, but photosynthesis cannot use the entire spectrum of incoming solar radiation ranging from ultraviolet through visible to infrared. Light absorbed by plant pigments is limited only to the visible segment of the spectrum—that is, to wavelengths between 400 and 740 nanometers, from violet to red light.[15]

This photosynthetically active radiation amounts to less than half (48.7 percent to be exact) of all incoming energy. Moreover, with green being the dominant color of vegetation, much of the radiation in that part of the visible spectrum is reflected: this amounts to about 10 percent of photosynthetically active radiation (4.9 percent of all incoming radiation), leaving nearly 44 percent (in the blue and red part of the spectrum) available to drive the conversion. Blue photons carry 75 percent more energy than red ones, but as this energy cannot be used rapidly or stored, some of it is lost, reducing the amount available to drive photosynthesis by the equivalent of nearly 7 percent of all incoming radiation. At this point in the sequence, we are left with about 37 percent of the solar irradiance available to energize photosynthesis.

Now we must subtract the energy needed to synthesize, and to regenerate, the compounds required to turn CO_2 into glucose. These reactions consume nearly 25 percent of the energy in the incoming solar radiation. And so, the best possible theoretical efficiency of carbohydrate photosynthesis is 12.6 percent. But we are not done yet. In most food and feed crops, the next reduction comes thanks to photorespiration—the reverse of photosynthesis—during which a significant share of newly produced phytomass is consumed, greatly reducing the overall net efficiency of the photosynthetic process. As mentioned, the enzyme Rubisco helps turn CO_2 into sugar (as a carboxylase), but Rubisco (as an oxygenase) can also act in the very opposite fashion, enabling the oxidation of newly made photosynthates and releasing CO_2.[16]

This loss depends on temperature and on the concentration of CO_2, and given the current atmospheric levels of O_2 and CO_2 (20.94 percent and 0.042 percent, respectively) it reduces the efficiency by an additional 50 percent, to only about 6.5 percent of the incoming radiation. In any case, without any changes in the cycle's biochemistry, only much-elevated concentrations of atmospheric CO_2, or a much-reduced presence of oxygen (to levels making animal life impossible), could eliminate this respiratory loss experienced by most (but as we will soon see, not by all) food and feed crops.

And now for the final efficiency deduction—obligatory, unavoidable respiration, a process quite distinct from undesirable photorespiration. This set of reactions is needed to ensure crop structure and function—that is, to supply energy for the synthesis of biopolymers from simpler compounds (cellulose from glucose, proteins from amino acids), for distributing the product of photosynthesis inside plants, and for repairing damaged or diseased tissues.[17] This generally increases with plant age, reducing net photosynthesis by as much as 75 percent or even 85 percent in some aged ecosystems. It remains moderate in crops that take only a few months to mature but it still claims about 30 percent of what is left after carbohydrate synthesis and photorespiration losses, reducing the final theoretically possible (maximum) photosynthetic efficiency—at the optimum temperature of 30°C and today's atmospheric CO_2 concentration (about 420 ppm)—to a measly 4.6 percent.[18]

Another pathway

Photorespiration reduces the photosynthetic rate in the two dominant staple grain crops, wheat and rice, as well as such traditional minor grains as rye, oats, and barley, all roots and tubers, and all leguminous and oil crops. But four important food crops—corn, sorghum, millet, and sugarcane—do not experience these losses because their photosynthesis follows a different pathway, first identified by Melvin Calvin and Andrew Benson in the early 1950s. Their discovery was made possible by labeling CO_2 molecules with a long-lived (heavy) isotope of carbon (^{14}C), and letting the photosynthesis proceed for anywhere between a fraction of a second and a few minutes before stopping its progress by alcohol and identifying the synthesized compounds.[19] That is how they discovered PGA—a compound containing three carbons ($C_3H_7O_7P$)—as the first stable product of the photosynthetic process, and hence the plants following what became known as the Calvin–Benson cycle are called C_3 species (they include all but a few major crops).

But when Hugo Kortschak, at that time working for the Hawaiian Sugar Planters' Association in Honolulu, repeated the investigation a few years later with sugarcane, he found that the first stable product was not the three-carbon PGA but four-carbon molecules of malate and aspartate.[20] In 1966, Hal Hatch and Roger Slack at the Colonial Sugar Refining Company in Queensland described the complete sequence of this alternative photosynthetic pathway.[21]

C_4 plants have two kinds of cells, one of them with lower oxygen and higher CO_2 concentrations, an adaptation that nearly eliminates all photorespiration, and they also have optimum growing temperatures 15–25°C higher than do C_3 plants. As a result, the best theoretical growing-season efficiency is about 6 percent—30 percent above the C_3 maximum (4.6 percent).

Actual field performance, even under optimal growing conditions, will be somewhat lower: measured short-term (hours, a day), photosynthetic efficiency is up to 3.5 percent for C_3 and 4.3 percent for C_4 plants—about 30 percent of the theoretical number. As already shown by the Manitoba wheat example, large-scale (regional, national) efficiencies calculated for an entire growing season will have to be much lower still. To begin with, incoming radiation is wasted before a sufficient leaf area has developed to intercept most of it. With spring wheat, it takes about 12 days after germination before the small plants have two leaves and about six weeks before the emergence of the last leaf. Before the plant canopies close, the incoming radiation is absorbed by soil, and variable shares of radiation continue to be lost even after canopies form and intercept the incoming sunlight—due to reflection by leaves and stalks and light transmission through canopies.

Sorting the wheat from the chaff

Another obvious detraction arises from the fact that, except for such leafy green vegetables as lettuces or spinach, we eat only a portion of the harvested phytomass, either because the rest is inedible or because we find it less palatable. In modern wheat and rice varieties, about half of the phytomass is straw (dry stalks and leaves), composed mostly of cellulose and hence not digestible by people. Other harvested parts may be edible (albeit less easy to digest), but most people prefer not to eat them: that is why we mill cereals. And so to calculate photosynthetic efficiencies per unit of *edible* yield, we must take our calculations a step further.

For staple cereals, the ratio of edible to inedible phytomass is usually expressed in terms of the harvest index. In cereals, the index is called the grain-to-straw ratio, and in traditional varieties of wheat, rye, oats, and barley these values were much lower than with today's short-stalked cultivars.[22] For wheat it was as low as 0.25 and no more than 0.3 (straw mass being three to four times grain mass);

for rice usually less than 0.35. These tall traditional varieties can be seen in many famous paintings. Joachim Patinir's *Rest on the Flight to Egypt* (c.1520), Pieter Bruegel the Elder's 1565 painting of harvesters, and Pieter van der Heyden's 1570 woodcut *Summer* show men harvesting crops reaching to, or even above, their shoulders.[23]

This means that the plants had heights of 1.3–1.5 meters. Of course, harvested straw was not wasted; it was used as animal feed and bedding, for thatching, raincoats, and sandals—the list goes on. In contrast, the modern short-stalked cereal varieties that have been dominant since the early 1960s are just 50–70 centimeters high. Their harvest index is 0.45–0.5 for semidwarf wheat (equal mass of harvested grain and straw) and close to 0.6 for paddy rice. This redistribution in what the photosynthesis is fueling—rather than any improvement in the overall efficiency of photosynthesis—has,

Pieter van der Heyden's 1570 woodcut (after Bruegel's drawing) shows men harvesting crops reaching to, or even above, their shoulders.

together with fertilization, been a principal cause of improved harvests.[24]

Moreover, as explained in the preceding chapter, most people prefer not to eat whole staple grains—that is, using whole-wheat flour in baking, or steaming whole grain, brown, rice. Bran makes up 14.5 percent and germ 2.6 percent of the whole grain, leaving 83 percent in endosperm, but most milled flours contain only 72–76 percent of the entire grain. Expressing the efficiency of Manitoba's spring wheat photosynthesis in terms of milled bread flour (75 percent extraction) would reduce the rate to only about 0.2 percent. Among C_3 crops, no photosynthetic efficiency rates—when extended to the entire growing season and no matter if in temperate or tropical environments—surpass 1 percent, but oil crops, potatoes, and rice do better than wheat.

Top of the crops

Rapeseed (canola) is now the leading oilseed crop, with the 2020 area nearly 40 percent larger than in the year 2000 and large swathes of farmland in North America, Europe, and China turning yellow during the weeks of flowering.[25] Rapeseed's photosynthetic conversion efficiency for the entire seed is 0.33 percent; and with oil content of 45 percent, this translates to a maximum efficiency of 0.15 percent in terms of edible oil, but the nutritious oil cake left behind after oil extraction is used as high-protein feed for animals, thereby contributing to additional food production.

For Manitoba potatoes, photosynthetic efficiency is about 0.6 percent. This is considerably higher than for wheat or canola, but the nutritional quality is far lower: a potato is almost pure carbohydrate, with a mere trace of protein and no fat, while durum wheat is sought after because of its high protein content, and canola is now the world's leading source of desirable polyunsaturated edible oil.

Because of their high yields, Chinese or Japanese rice varieties (irrigated and heavily fertilized) do better than Canadian spring wheat. No less than 0.4 percent of solar radiation reaching a wet field in Jiangsu during the crop's 120-day growing season is converted into harvested grain (6 tons per hectare). Depending on the rice variety and the milling process, milled white rice contains only 60–65 percent of the whole grain's mass, and this reduces the photosynthetic efficiency to no more than 0.25 percent.[26]

As expected, C_4 crops—those without the largely wasteful photorespiration process—do even better. The worldwide harvest of corn is now nearly as large as the combined harvest of wheat and rice, but while those two grains are used almost solely for food, different varieties of corn are used as animal feed in the US (accounting for 99 percent of the harvest—the world's largest—with food-grade corn being the remaining tiny 1 percent), as well as in the EU and China. Corn is used as food in Mexico, where the crop was domesticated, and widely throughout sub-Saharan Africa, where it is a staple for more than 300 million people and provides about a third of food energy intake in the form of cornmeal (ugali).[27] And in the US, grain corn is also used to produce automotive ethanol.[28]

More than half of a mature plant's above-ground phytomass is accounted for by the corn stover (dominated by stalks and leaves, and including cobs, sheaths, husk, and tassels and silks). The stover is indigestible by humans, but it is digestible by ruminants. The production of harvested phytomass (stover plus grain) proceeds with much higher photosynthetic efficiency than the growth of corn grain alone, be it fed to animals, nixtamalized (soaked and cooked in alkaline solution) to make tortillas, or ground and cooked to make ugali. Record yields of US corn are produced with the highest photosynthetic efficiency.

In 2020, optimally fertilized Iowa corn (178 bushels per acre, or 11.1 tons per hectare) reached a solar conversion efficiency of 0.7 percent. There are no further subtractions, as the grain is fed (directly, or after milling and mixing with other feeds) to animals. In contrast, sweetcorn (sugar corn) grown for direct human consumption (and classified by the US Department of Agriculture as a vegetable) yields nearly 2.5 times (about 25 tons per hectare) as much in Washington (the state with the largest production), but its high water and low energy content (3.6 megajoules per kilogram, compared to about 16 MJ/kg for grain corn) translates into a photosynthetic conversion efficiency about 50 percent lower (0.35 percent).[29]

And the winner is . . .

This should leave sugarcane as the undisputed top photosynthetic efficiency performer. Surprisingly, that is true only as far as the total above-ground phytomass production is concerned. Sugarcane is a perennial grass that is left to regrow four or five times a year before replanting. Grown in a tropical climate and supplied with adequate water, this leading C_4 crop also benefits from its association with nitrogen-fixing bacteria (they live within its stalks and leaves and provide a significant share of its macronutrient needs), and its record small-scale yields can surpass 100 tons per hectare.

Freshly cut sugarcane stalks are mostly water (63–73 percent), with fiber and sugar accounting for similar shares (12–16 percent) of the total phytomass. With energy content of about 7.2 gigajoules per ton and with solar radiation hitting a horizontal surface during the 12 months of growth at 62 terajoules (typical of São Paulo, Brazil's leading cane-producing state), this would imply an efficiency of 1.2 percent. Average yields are lower (in 2019–20 the Brazilian mean was 76.13 terajoules per hectare), and to get the food efficiency value corresponding to actual edible food output (that is, the equivalent of wheat flour or rapeseed oil) we must consider only the total recoverable sugar. In 2019–20 it averaged 139.3 kilograms per ton of fresh-cut stalks—that is, 10.6 tons of sugar per hectare and a photosynthetic conversion efficiency (with 16.6 MJ/kg) of no more than 0.28 percent.[30]

This means that in terms of edible energy, sugarcane's photosynthetic efficiency is only slightly better than that of milled East Asian rice and lower than that of potatoes or American sweetcorn. And if we compare the nutritional qualities of these foods then sweetcorn looks even better. Of course, it is mostly (73 percent) water, but it contains small amounts of proteins and oil while cane produces only pure sugar. Similarly, the slightly lower photosynthetic energy efficiency of producing milled rice is easily compensated by the grain's superior nutritional quality (raw rice is 7.5 percent of protein and is relatively high in the micronutrients phosphorus and potassium).

Modern farming practices and better crop varieties have improved energetic efficiencies, but these gains have not been the result of any fundamental improvements in the process of photosynthesis. Instead, they followed an increased supply of water (irrigation; conservation tillage to store more rainfall in soil) and nutrients (fertilization), as well as the introduction of new crop varieties that were able to take full advantage of these optimized supplies and were selected in order to redistribute part of the newly synthesized phytomass from a plant's inedible to digestible parts.

Even so, these gains in photosynthetic efficiency have been relatively limited, and more substantial improvements could come only with redesigning the photosynthetic process, which would necessarily be extremely challenging. The first experimental steps toward this goal have been taken, but given the complexity of the process we should not expect a stunning transformation anytime soon; food production will remain inherently inefficient in its use of sunlight for a long time to come. Fortunately, the amount of sunlight striking the earth does not limit food production.

Water and nutrients

In comparison to the low efficiency of energy conversion, other essential inputs into the photosynthetic process—water and plant macro- and micronutrients—are processed with significantly higher efficiencies, but in absolute terms their losses remain unacceptably high, adding both to the cost of production and to the environmental impacts of crop cultivation. So here let's quantify these inherent inefficiencies by focusing on water losses and on the use of nitrogen—the most important (both in terms of its total mass requirements and its subsequent environmental impacts) macronutrient, whose supply is often decisive in ensuring the best possible crop yields. Only then is it possible to explain the best practical expectations and list some effective ways to reduce the losses in crop production—rather than engaging in fact-free speculations.

Inefficient water use in photosynthesis is easy to appreciate once one understands what plants must do to import water and CO_2 into their leaves. Soil water is absorbed through the roots, then transported to leaves, where it is required for the photosynthesis of new plant mass. However, far more of it is lost through stomata, tiny openings usually sited at the underside of leaves. Their two guard cells open or close the pores to let CO_2 in and water and oxygen (released by photosynthesis) out.[31] This process of transpiration (essentially pulling water from soil) results in a grossly lopsided exchange of water for carbon. Fresh leaves contain anywhere between 70 percent and more than 90 percent of water, while the typical atmospheric water content ranges from 1–2 percent in temperate regions to about 4 percent in the tropics, and CO_2 makes up only 0.04 percent of air.

This means that living leaf cells are water-saturated while the air is considerably undersaturated, and this difference produces a one-way flow transferring water from soil to roots to leaves to the air. Because the difference between water vapor pressure inside the leaves and in the atmosphere is two orders of magnitude higher than the difference between the internal and external vapor pressures of CO_2, C_3 crops need between 400 and 1,600 grams of water in order to incorporate a gram of carbon in new phytomass, while more water-efficient C_4 plants can manage with "just" 160–250 grams.[32] Because of these high rates, water shortages rather than insufficient energy supply (via the light that drives photosynthesis) are the most common limiting factor in crop production (they affect about two-fifths of all cropland), and crop production makes by far the largest claim on human use of fresh water resources, accounting for about 80 percent of all water used.

Quantifying water losses

Vapor pressure deficit (VPD) measures the difference in vapor pressure inside and outside a leaf, and it varies with temperature and humidity.[33] When the VPD is low, transpiration slows down or it may even cease. In very dry climates with a very high VPD, plants would close the stomata in order to prevent leaves from drying out and to maintain their essential moisture. But that reduces the flow of nutrients absorbed by the roots and cuts off the inflow of CO_2, inevitably slowing down photosynthesis and inhibiting growth. The regulation of transpiration is a key process that determines the rate of plant growth under a high VPD, and plant scientists measure a crop's water use efficiency (WUE) in terms of new plant mass photosynthesized per unit volume of water used by a plant, as they have done since 1913.[34]

As expected, there are notable regional variations. The WUE for wheat, defined from the short-term gas exchange of CO_2 and H_2O, ranges from 29 to 105 kilograms per hectare-millimeter for all above-ground biomass, and from 5.4 to 24 kilograms per hectare-millimeter for grain yield—with ranges from 9.9 in southeastern Australia and 9.8 on China's Loess Plateau to 7 for the Mediterranean Basin and just 5.3 for the southern-central Great Plains, compared to the theoretical maxima of just over 20.[35] Most of this variation is accounted for by water evaporation around the time of wheat flowering, with additional contributions due to phosphorus deficiency, late sowing, poor soil quality (alkaline, saline), crop disease, weeds, and lodging (the bending of stems near the ground).

But I admit, hectare-millimeter as a denominator is not an easy unit to comprehend, and so for our purposes perhaps a more helpful way of comparing WUE rates among crops is an inverse rate indicating how much water is required per unit mass of crop yield. People who have never encountered these rates before will find them stunningly large.[36] The average amount of water required to produce staple cereals is about 1,600 tons per ton, with wheat around 1,800 and corn at "just" 1,200. In volume terms, a ton of water fits into a 1-meter cube; 1,600 tons of water fits into a cube with sides about 11.7 meters long, or the height of a typical two-story American house. Legumes have even higher water demand, averaging around 4,000 t/t, the rate topped by vegetable oils—from soybean at 4,200, peanut at 7,500, and olive at nearly 15,000 t/t, and nuts (9,000 t/t), and even more so by coffee.

With about 18,000 t/t, coffee's water use prorates to about 130 liters (kilograms) per cup (about 250 milliliters) made with 7 grams of roasted beans! Sugar crops are most water-efficient (less than 200 t/t), and the rates for fruits and vegetables range from below 300 (watermelons, pineapples, plums) to 800–1,000 (bananas, apples, pears, peaches) and peak above 2,000 (grapes, dates, figs). Beer needs about 300 t/t, wine nearly three times as much, and orange and apple juices more than wine (1,000–1,100 t/t).[37]

All else being equal, the water use efficiency of plants is almost directly proportional to the atmospheric concentration of CO_2. That is why during the last ice age, when CO_2 levels were just 180 parts per million, plants had to transpire twice as much water as during the last decade of the twentieth century, when the level rose to 360 ppm—and (emphasizing again: everything else being equal) rising concentrations of atmospheric CO_2 should improve WUE.[38] Experimental results support this expectation: with about 700 ppm CO_2, most C_3 species should yield nearly 30 percent more, and long-term global trends suggest the WUE of grain cultivation is increasing.[39]

Deliberate CO_2 enrichment has been used for decades to reach full photosynthetic potential, by lowering the water use and increasing the yield of crops grown in greenhouses. Dutch producers have been leaders in this field, enriching primarily their two leading vegetable crops—peppers and tomatoes—as well as the growth of cut flowers.[40] They rely on cogeneration, burning natural gas to produce electricity and heat for the greenhouses and using a part of the generated CO_2 to enrich the enclosed air spaces to at least 1,000 ppm—nearly 2.5 times as much as the current atmospheric level. But it is important to note that while rising CO_2 concentrations can ease water stress, they cannot help with heat stress. As a result, in a warmer world, some heat-tolerant species would yield more even with reduced precipitation, while other species would have their yields reduced even with adequate water supply. Global warming will have some positive and many negative effects, all of which will display major regional variation, and their overall impact on crop yields, as we will see, could be reduced by better agronomic practices.

Nitrogen: the most important macronutrient

And there is yet another major character in the story of photosynthetic inefficiencies. You will recall that the only reactants in the simplest version of the photosynthetic reaction repeatedly used in biology textbooks are carbon dioxide and water. But, as already explained, no phytomass-producing reaction can proceed in the absence of the requisite macro- and micronutrients. Nitrogen is the macronutrient that is needed in the relatively largest amounts and whose deficiencies also put the most common limit on higher crop yields.[41] Using, once again, spring wheat as a representative example, a good Manitoba crop of this staple cereal (3 tons per hectare) needs 87 kilograms of nitrogen but only about 11 kilograms of phosphorus and potassium.[42] And, unfortunately, nitrogen's compounds—applied as synthetic fertilizers, or present (or converted from) recycled crop residues and manures—are especially prone to processes or reactions through which they are lost to water, air, and soil before getting a chance to be absorbed by crop roots.[43]

Given the fact that nitrogen is the most commonly deficient macronutrient, that nitrogenous fertilizers usually account for the single largest share of variable costs in intensive crop farming, and that lost nitrogenous compounds have highly undesirable environmental impacts, it is not surprising that the best field operations try to minimize nitrogen loss. It is also unsurprising that we have many studies of nitrogen use efficiency (NUE) in crop cultivation: this is, simply, the quotient of nitrogen output in harvested crop removed from the land and the nitrogen inputs into crop cultivation (or into specific livestock products).[44]

NUE can also refer to all nitrogen available to a growing crop (in soil organic matter, in recycled crop residues, in added animal manures and fertilizers), but most research has been devoted to tracing the transfer of the most massive, and the most expensive, input—that is, the share of applied nitrogenous synthetic compounds (urea, ammonia, nitrates) that ends up in harvested plant parts.

These applications range widely, from more than 200 kilograms per hectare for China's double-cropped rice to just a few kilograms of urea (solid nitrogenous fertilizer made from ammonia) broadcast by hand by African farmers on their small plots. Typical rates in European wheat-cropping are around 100 kg/ha, and in the 120–150 kg/ha range for Indian rice.[45] Grain corn usually has the highest NUE among staple grains (globally at 33 percent; two-thirds of applied nitrogen do not reach the plants) and rice the lowest (often just above 20 percent). Given the size of the globally fertilized area and differences in climate and in farming practices, it must be expected that large-scale studies of NUE will come with a range of results. A global high-resolution assessment of nitrogen flows in croplands in the year 2000 concluded that about 35 percent of the available macronutrient was taken up by harvested crops and 20 percent by their residues, with the remainder lost in leaching (16 percent), soil erosion (15 percent), and gaseous emissions (14 percent).[46]

In 2013, a study tracing the global and regional trends in nitrogen efficiency between the 1960s and 2007 found the overall recovery of the nutrient by harvested crops was "about the same at the beginning of the time series as it was at the end"—averaging about 40 percent.[47] As expected, efficiencies were higher in high-income countries, and there was a great deal of variation in NUE in low-income economies. While Brazil, India, and the USSR/Russia had recorded increased recoveries, the Chinese mean declined from 37 percent to 29 percent. In 2014, an evaluation of global NUE came up with different values, showing a decline from as much as 69 percent in the early 1960s to just 45 percent by 1980, followed by a slight rise and stabilization at about 47 percent.[48] Within Europe, evaluation of monetary returns created per unit of nitrogen stock showed Italy, Spain, and Austria to be the most efficient countries, and Ireland, the UK, and Norway the least efficient ones—but this rating is not directly comparable with standard NUE, as it is greatly affected by the value of crops produced (grapes vs. grain; fruits vs. legumes).[49]

This chapter has been full of numbers and simple calculations for a simple reason: there is no better way to appreciate both the surprisingly low efficiency of photosynthesis and its high demands for its principal material inputs of water and nitrogen. The next chapter will use a similar approach as we unpick why we eat so many of certain animals and not others.

4. Why Do We Eat Some Animals and Not Others?

The statistics are clear: an overwhelming majority of this book's readers are not vegetarians (in affluent countries, no more than about 5 percent are), and this means that they must make repeated choices about what kind of meats (and in what quantities) to consume.[1] But no matter if you have clear dietary preferences (chicken above beef) or if you are omni-carnivorous, chances are that you have never asked some fundamental questions concerning meat consumption. Why have we domesticated such a small number of animals to produce meat (and milk and eggs and wool), and to help us in transportation and other work? As we will see, we chose the limited number of animals that we did because of a cocktail of considerations including size, metabolism, social organization, behavior, and feeding habits (or, as ecologists would say, trophic levels).

Perhaps the best place to start is to ask: what species of animals would our ancestors have preferred for domestication if they had today's best scientific understanding regarding animal metabolism and trophic levels? The answer is exactly the same animals as they did select, starting about 11,000 years ago with, in fairly quick succession, goats and sheep, followed by pigs (10,500 years ago), and cattle (10,000 years ago), all within a crescent-shaped area in what is now southern and eastern Turkey, northern Iraq, and northwestern Iran.[2] Other domestications came considerably later: donkeys and yaks about 7,000 years ago, water buffalo and camels 6,000 years ago, llamas and alpacas 5,500 years ago, and horses only around 2,500 BCE.[3] The history of food production has a way of selecting the best methods, and this shouldn't be surprising: long periods of observation, experience, and trials were

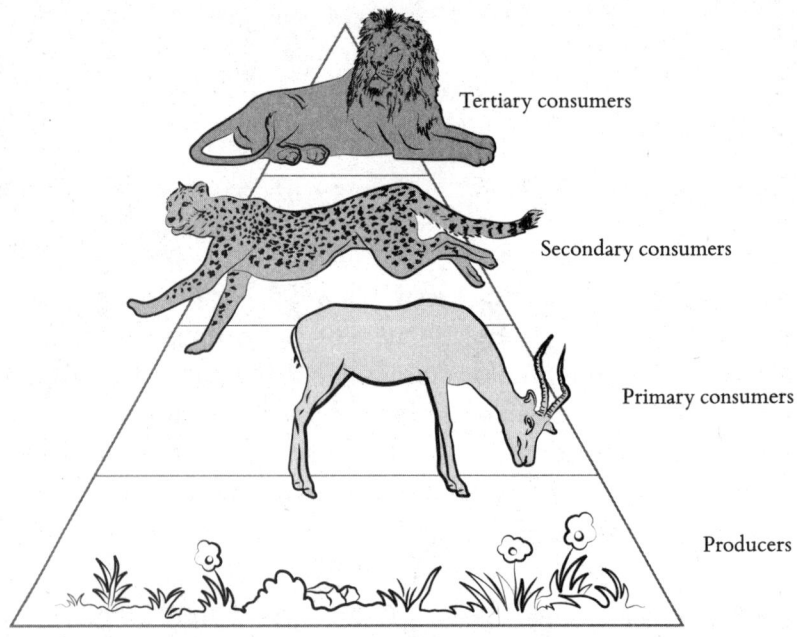

Trophic levels: the terrestrial pyramid of life.

bound to end with the same outcomes as those dictated by formal scientific understanding.

Quantitative clues

The biosphere contains nearly 6,500 mammalian and almost 10,000 avian species, and yet the Food and Agriculture Organization's annual statistics count only 16 categories of domesticated animals.[4] Most are single species (with numerous varieties); three categories combine two similar species (camels: dromedary and Bactrian; and camelids: llamas and alpacas); two contain two different species (rabbits and hares; geese and guinea fowl). Of these 16 groups, 12 are mammals and 4 are birds, and their total numbers are distributed in a highly skewed manner. Larger ruminants dominate the mammalian count with more than 4 billion heads: about 1.5 billion cattle, 1.25 billion sheep, 1.1 billion goats, some 200 million water buffalo, and 40 million camels.[5] Pigs rank next, at nearly a billion; horses, asses, mules, and camels add up to only about 130 million; and in 2020 there were fewer than 200,000 caged rabbits (and hares) and fewer than 20,000 domesticated rodents (mostly Andean guinea pigs). Among birds, chickens—after a rapid global diffusion—are vastly more numerous than any other domesticated avian species: about 33 billion in 2020, compared to some 1.1 billion ducks, and less than half a billion each of turkeys and geese.[6]

All of these animals are herbivores, although pigs, as well as ducks and geese, are opportunistically omnivorous. Less than 5 percent of a wild pig's diet consists of insects and small animals, and domesticated pigs will feed on garbage of any kind or on carcasses of animals, while ducks will eat, if they come across them, insects, small fish, and crustaceans, and so would geese, but both species are overwhelmingly herbivorous.[7] The biosphere's macroscopic pyramid of life—which you might remember, however vaguely, from school—rests on the photosynthetic productivity of plants (primary producers, the first trophic level), and the most abundant mammals will be herbivores (primary consumers at the second trophic level) able to feed directly on various plant parts.[8] The number of secondary consumers (the third trophic level, including foxes and bears)—carnivores eating primary consumers and omnivores eating plants and primary consumers—and their total zoomass will be necessarily smaller.

Tertiary consumers (the fourth trophic level) can also be omnivorous, but the carnivorous species eat both primary and secondary consumers: lions will kill antelopes but they will also kill and, when hungry, eat cheetahs (although, there are exceptions to everything, and they do not eat the leopards or hyenas that they kill). Tertiary consumers are much more common in the ocean, with sharks and tuna the most massive examples. The aggregate zoomass of all bluefin tuna in the ocean is much smaller than that of the carnivorous mackerel (secondary consumers) on which they feed and vastly smaller than that of herring (primary consumers), which are also their common prey.

Not only are the most numerous domesticated mammals—cattle, sheep, and goats—herbivorous, they are ruminants, capable of digesting abundantly available lignocellulosic phytomass (see the first chapter), and hence have at their disposal vastly more biomass than most of the mammals who are unable to digest those compounds. But being a herbivore (and especially a ruminant) is not enough to be selected for domestication: size and behavior are both very important.

Size matters

Size matters in two important ways: in feeding animals and in taking care of them. The specific basal metabolism (the energy required at rest per unit of body weight) of warm-blooded animals decreases with increasing body size.[9] A 400-kilogram cow will need only about 60 percent as much energy per kilogram of body mass as a 40-kilogram sheep, and the sheep will need only about 25 percent as much energy as a 400-gram rat, and the rat has a far lower specific metabolism than a 20-gram mouse.

So, even if you were to confine mice in very small cages (and hence reduce their energy needs close to their basal or resting metabolism), they would still need almost five times as much feed per unit of weight as a small goat, nearly ten times as much as a large pig—and all that to get too much skin (smaller animals have relatively larger body surface areas) and mere grams of meat per animal. This is why mice were caught and eaten by some societies (Romans ate dormice, *glires*, fattened in jars, gutted, and stuffed with minced pork) but were never really domesticated.[10] And this is also why the annual meat output from the smallest domesticated mammals—guinea pigs (they need only about a third as much feed per unit of body mass as a mouse) and rabbits—has remained limited.

Guinea pigs (yielding less than a kilogram of meat per carcass) have never made it globally as a source of food and their rearing remains confined to its original region of domestication as they are fed food scraps in Andean kitchens.[11] Rabbits are larger, with three-month-old animals having live weight mostly between 1.5 and 2 kilograms, and heavier (up to six months old) rabbits for roasting weighing 2.5–3.5 kilograms. With carcass weight being about 50 percent of the live weight, a small rabbit yields less than a kilogram of meat, still too small to become a global favorite: their production is highly concentrated in just a handful of countries dominated by China and including North Korea, Egypt, Italy, and Russia.[12]

What's for dinner?

As you can see, metabolic considerations and typical meat yields favor larger mammals—animals found in plentiful numbers whose carcasses provide enough meat for at least an entire family. Goats and sheep have all the desirable attributes facilitating domestication.[13] They are the right sizes: most adult sheep weigh between 50 and 120 kilograms, and larger varieties of adult goats can match that range. Small goats, common in parts of Asia, weigh only 25–40 kilograms. Both species are ruminants and are able to adapt to harsh climates (arid, cold), and goats, in particular, eat thorny plants and climb steeply inclined surfaces with their split hooves and rubber-like soles. And their suitability for domestication is further enhanced by their social behavior. Sheep have an intensely gregarious social instinct resulting in a strong flock mentality, and goat flocks can even mingle with sheep, cows, and horses.[14]

While Islam, Hinduism, and Judaism label pigs as unclean animals, a proscription that reduces the pool of potential pork consumers by about 3 billion people—or nearly two-fifths of humanity—pork is still by far the most consumed mammalian meat.[15] About 110 million tons of it was produced in 2020, which is about 60 percent more than beef and 3.5 times as much as goat and sheep meat combined. As with sheep and goats, a combination of attributes facilitated the domestication of wild pigs. Mass-wise they are compromise mammals (not too small, not too large) and their body mass range overlaps our body weights (mostly between 60 and 150 kilograms). Pigs are also adaptable to a range of climates (from the tropics to the sub-Arctic), their omnivory makes them easy to feed, they are social animals that can be herded like sheep, and like sheep (and unlike goats), fats make up a relatively high share of their carcasses—a welcome attribute in societies whose traditional diets were short of fats.[16]

Cattle, much like sheep and goats, are ruminants (more grazers than browsers, preferring grasses to leaves and branches) with distinct social structures.[17] They recognize human faces, and when properly handled they are docile and readily herded. But they weigh considerably more than sheep, goats, and pigs, which has the advantage of providing plenty of meat per animal. Although their heavy bones and skin reduce their edible carcass weight to only about 40 percent of live weight (for pigs this ratio is about 60 percent), their large size means that a mature animal will yield at least 150 kilograms (for low-weight Indian cows) and up to 350 kilograms of meat for large European breeds.

But this high meat yield per animal carries high metabolic costs: to reach such substantial slaughter weights takes long periods of feeding, and although the basal metabolism of larger mammals is lower than that of smaller species, this advantage is negated by their much larger body mass: cows of heavy cattle breeds can weigh more than 700 kilograms; the heaviest bulls can weigh more than a ton.[18] As a result, even modern meat breeds, fed carefully balanced mixtures in large feedlots, go to slaughter only after at least two years of this practice. We will assess the energy requirements of specific feeding regimes later in this chapter, but for now, it is clear that producing beef is less efficient and more environmentally demanding than producing pork.

Cash cows

So why did our ancestors domesticate not only smaller (mid-weight) species but also large—and in terms of meat production, very expensive—cattle, water buffalo, and camels? Because they did not do so for meat, but for their muscular power (oxen, castrated bulls, were used as draft animals for farming and for road transport) and for being daily providers of the best-quality protein (in milk). For centuries, oxen remained the dominant draft animals not only in Asia and Africa but also in Europe—and, following European colonization, in the Americas.[19] Horses began to make inroads in parts of Europe starting in the High Middle Ages, but in some regions, oxen remained indispensable well into the 19th century. For example, in the early 1880s oxen accounted for more than 60 percent of all draft animals in western and southwestern France, and their retreat in Europe and the US was connected to the introduction of new grain-harvesting machinery (reapers, binders, later combines), designed to be pulled by more powerful horses.[20] And as long as oxen were needed for essential field work, they were killed for meat only when aged, ill, or when feed became scarce.[21]

The importance of milk and dairy products in ancient, medieval, and early modern diets in regions with domesticated cattle is attested to in numerous ways. A Roman breakfast (*jentaculum*) was commonly bread and cheese, and some two millennia later Frederick Morton Eden's inquiry into the lives of the English poor (published in 1797) described simple breakfasts of boiled cereal or leguminous grains, usually taken with "a little milk," providing a small, but daily, dose of high-quality protein.[22] And while the central place of the cow and milk in Hinduism remains as strong as ever (India has nearly 12 percent of the world's cattle, nearly 55 percent of all water buffalo, and is the world's second-largest milk producer, close behind the US), milk has gained importance even in some traditionally non-milking civilizations, with Japan being the foremost example. In recent years the average annual per capita consumption of dairy products (milk, yogurt, ice cream, and cheeses) has been surpassing the weight (but not the total food energy value) of rice.[23]

Beef was always eaten in Europe (and also in parts of Asia and Africa), sometimes in substantial amounts, by those who could afford it, but three major shifts had to happen to transform it into one of the world's leading meats. First, the displacement of cattle as the most important source of traction in farming.[24] Second, the rise of beef raised on the vast natural pastures of the United States, Canada, Argentina, and Australia, which took place during the second half of the 19th century and led to the first large-scale intercontinental exports of beef. And, finally, the displacement of horses by internal combustion-powered machines (tractors, combines) that made it possible to use large shares of land previously devoted to horse feed (as much as 25 percent of all US farmland during the second decade of the 20th century) for producing concentrate feeds (mainly corn and soybeans), whose availability made it possible to produce unprecedented quantities of beef.[25]

US beef production data illustrate the result of these changes. In 1900 the country produced about 2.5 million tons of beef, and that total had grown only marginally by 1930.[26] Requirements for military beef rations (canned C-rations contained mostly corned beef) raised the output to 4.7 million tons by 1945, and then beef production nearly tripled between 1950 and 1976 (to 11.8 million tons) as concentrated animal feeding operations (CAFOs) became the dominant source of inexpensive beef. The subsequent stagnation of output was due to concerns about the health effects of beef-eating (particularly its supposed role in rising cardiovascular mortality), and new, only slightly elevated records (implying considerably lower per capita production) were set only in the year 2000 (12.2 million tons) and in 2020 (12.3 million tons).[27] As post-1950 beef production increased in Europe and in Latin America, the US share of world output has been declining steadily, from more than 30 percent in 1950 to 20 percent by 2020.

Animal farms

But beef's relative retreat did not mean lower per capita meat consumption, be it in the US or worldwide; it was compensated for by the rise of pork and by the even faster rise of chicken meat. In 2020 the FAO's global data listed about 120 million tons of chicken, 110 million tons of pork, and 68 million tons of beef. During the past 50 years, both pork and chicken meat output have risen thanks to the industrialization of operations such as CAFOs of increasing size replacing previously dominant (and, in comparison, always small-scale) mixed-farm production, where several species of animals were raised on the locally harvested feed. This traditional link was severed by new facilities housing more than 1,000 animal units (AU). The relative measure is based on feeding requirements, with a head of cattle ready for slaughter at 1.0, a dairy cow 1.4, a maturing pig of over 25 kilograms at 0.4, and a laying hen or chicken at 0.01 AU.[28]

Consequently, a cattle-feeding CAFO must have at least 800 heads, a pig-feeding one at least 2,500 hogs, and a broiler house more than 33,000 chickens. The largest American operations are much bigger, with beef lots accommodating more than 100,000 large animals.[29] All of these operations are just for a single species supplied with meals formulated for the fastest possible weight gain and shortest spells to reaching slaughter weight: mixtures of carbohydrate (dominated by corn) and protein (dominated by soybeans), cereal and leguminous grains (for beef this is supplemented with roughage). There are no opportunities for extended free movement and hence no vegetation, grazing, rooting, or perching.

CAFOs have been criticized, for decades, for many reasons. By the opponents of meat-eating for their cruelty to animals, by environmentalists for their numerous impacts on land, air, and water quality (generating pollutants, pathogens, and infestations), by ecological economists for their high resource-intensity that makes them unsustainable and economically questionable.[30] In many affluent countries, usually more expensive alternatives—ranging from grass-fed beef and free-roaming pork to free-run chicken and eggs—are now widely available; but in the US, large shares of commercially produced animals (70 percent of cows, 98 percent of pigs, and 99 percent of chickens and turkeys) now come from such facilities, and CAFOs currently supply more than 70 percent of the world's poultry and more than half of the world's pork.[31]

CAFOs attained such dominance by greatly shortening traditional feeding spans and by lowering the price of meat. In this respect, the most notable shift took place in the consumption of chicken. Before 1920 in the US (and before the 1950s in most EU countries), chicken meat was a luxury served at special occasions because the birds were kept for egg-laying, and their meat became available only as a result of culling cockerels and unproductive hens.

Commercial breeding and raising chickens for meat began to develop only during the 1920s in the US and after the Second World War in Europe. In modern CAFOs pigs go to market after just six months, and modern CAFO broilers (chickens raised for meat) are ready in just six weeks. This shortening of time before marketing, resulting from the combination of breeding, high-quality feed, and rearing the animals in confinement (reducing their energy metabolism), has produced the most remarkable efficiency gains in chicken, a much less pronounced improvement in pigs, and no gain in cattle.

Which meat is most energy-efficient?

Thanks to the US Department of Agriculture, we can trace average feeding efficiencies since 1910 for cattle and pigs, and since 1935 for chickens.[32] The rates are expressed in terms of feed units (equivalent to the energy content of grain corn) per unit of live weight (or per unit of milk, or per 100 eggs). Despite major changes in breeding and feeding, the intensity of beef production has shown plenty of fluctuation but no singular trend: in 1910 it took about 10

Why Do We Eat Some Animals and Not Others?

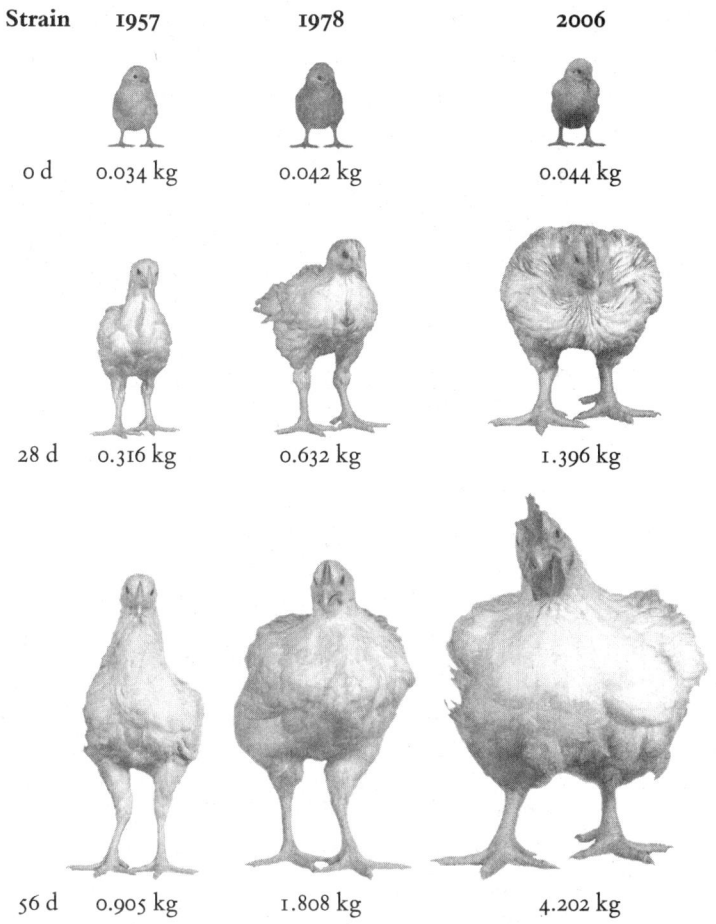

Strain	1957	1978	2006
0 d	0.034 kg	0.042 kg	0.044 kg
28 d	0.316 kg	0.632 kg	1.396 kg
56 d	0.905 kg	1.808 kg	4.202 kg

What modern feeding and breeding did to a chicken: front views of birds at 0, 28 and 56 days of age in 1957, 1978 and 2006.

kilograms of corn equivalent to produce a kilogram of live cattle weight; subsequent fluctuations raised the rate to about 14 by 1980; and the latest values (for the late 2010s) are about 12. Similarly, pork's feeding rate declined from 6.6 in 1910 to 5 by 1930, was at 6 by 1980, and recently it has been just below 5. Consequently, the only demonstrably impressive decline of feeding intensity has been chicken production, falling from 5.5 in 1925 to 4 by 1950, 3 by 1960, 2 by the year 2000, and 1.6 by 2010, followed by a slight increase.

These USDA rates are often used to conclude that pork takes about three times more feed energy to produce than chicken meat, and that beef production demands at least seven times more feed than raising broiler meat.

The quoted differences are correct, though they do not refer to meat but (as the USDA explicitly says) to a unit of live weight, and in order to make the rates truly comparable we must convert live weight to edible weight, a species-specific adjustment. As already noted, beef (with heavy bones and large skin area) has the lowest edible meat/live weight ratio of about 40 percent. In contrast, a plucked chicken can be eaten in its entirety in Asia (including feet and rooster's comb, while only beaks may be left), and even in the US (where it is now eaten mostly as cut-up portions: breasts, thighs, wings) the ratio is 60 percent, while for pork nearly 55 percent of the live weight is edible.[33] Adjusting for these specific rates we end up with actual recent US feeding intensities of about 2.7 for chicken meat, 9 for pork, and 30 for beef. If the choice of meat were based solely on the efficiency of feeding, then it would make no sense to eat beef, a meat that requires over 10 times more feed energy per unit of edible product than chicken.

At the same time, it must be remembered that these comparisons, done in terms of feeding units (corn equivalents), do not tell us anything about the actual composition of specific animal diets. While beef has a substantially higher feed energy cost than chicken, even in the US a large part of that feed comes from forages and crop residues that are not edible by humans, while chicken may be fed with a mixture of grain and soybeans that could have been turned directly into edible products. So far, the most comprehensive study of livestock feed has shown that 86 percent of it (measured in terms of dry matter) consisted of materials not edible by humans, and that to produce a kilogram of meat requires, on average, 2.8 kilograms of human-edible grain for ruminants and 3.2 for monogastric animals (pigs, poultry)—and that only modest gains in feed conversion efficiency would suffice to prevent further expansion of arable land devoted to animal feed (grains, oilseeds) production.[34]

And these feeding intensities also mean that the overall energy efficiency of the global food system, with increasing amounts of meat (as well as eggs and milk) consumption, has been declining. Recall that even a good harvest of Iowa corn (a relatively efficient C_4 crop) embodies only about 0.7 percent of the solar radiation received by the field, and that the efficiencies are lower (0.15–0.6 percent) for C_3 crops. Consequently, feedlot beef raised on corn would embody merely 0.023 percent (0.7/30) of the solar energy required to grow the feed, and a chicken fed a mixture of corn, wheat, and soybeans (assuming a weighted production efficiency of 0.35 percent) would contain just 0.13 percent of the energy that reached the fields used to grow those feedstuffs.

Inevitably, eating animal foodstuffs is further lowering the overall efficiency of our food production, and this reality—resulting from the consumption of foodstuffs that come from higher levels of the food chain—can also be expressed by tracing the shifts of average human trophic levels.

What's at stake?

Populations eating solely plant foodstuffs would be at a trophic level of 2.0 (recall the trophic pyramid shown earlier in this chapter), and many regions in premodern China and Africa came very close to pure plant diets. (Chinese dependence on grain staples was noted in the second chapter.) In contrast, people who would follow the so-called Paleolithic diet recently promoted by some dieticians ("steak and salad" might be its best brief definition) would be eating at a trophic level very close to 3.0.

In 2013 a group of French scientists at the Institut Français de Recherche pour l'Exploitation de la Mer published a global quantification of national and global trophic levels based on the FAO's food supply statistics, and found that between 1961 and 2009 the global median (weighted by the population size of each country) rose from 2.13 to 2.21—and that this increase was mainly driven by

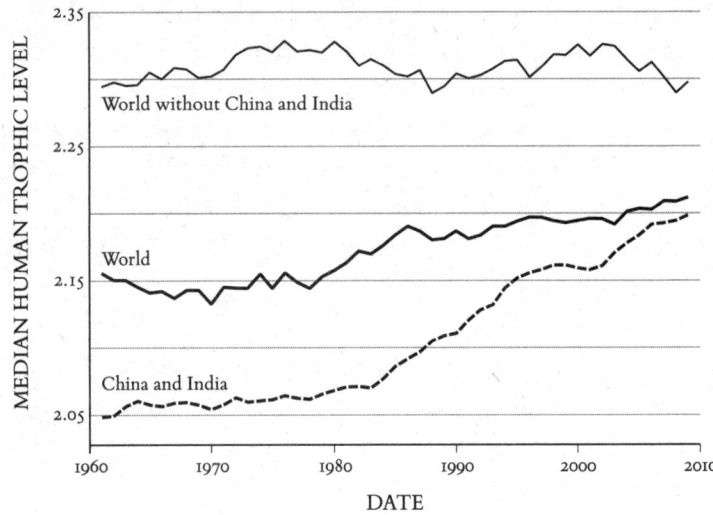

Median food trophic levels rising: the Asian effect.

dietary transformations in China and India, with their median human trophic level rising from 2.05 to 2.20, while the level for other countries remained stable at 2.31, with the Icelandic diet (half plants, half fish and meat) taking the top rating of 2.57, down from 2.76 in the early 1970s.[35]

Although this analysis confirmed the existence of a universal component of dietary transformations—eating at even a slightly higher average trophic level is putting a greater stress on the biosphere's resources—the authors did not point out a fundamental problem with their basic data: the FAO's food balance accounts are for the supply of food, not for its actual consumption. For example, the FAO's recent food balances for France show an average daily supply of 3,500 kcal/capita, but no more than 2,200–2,400 kcal/capita is eaten; the rest is accounted for by ubiquitous and unacceptably high food waste at all levels (production, processing, wholesale, retail, household). But as it is impossible to make any reliable estimates regarding the national and temporal differences in food losses (and hence to know the amounts of food wasted within specific categories), we have no way to adjust human trophic levels from supply to actual intake, whose level is only rarely measured in a direct and reliable way.

Greener pastures?

When we turn away from modern beef produced in massive feedlots and dependent on mixtures of corn and soybeans, and consider the meat produced solely by grazing, the trophic level remains the same (2.0) but the conversion efficiency comparison changes profoundly. If the number of animals in the pasture is carefully limited to the grazing area's carrying capacity (that is, to the amount of new plant mass that can be produced by a specific grassland and consumed by animals without causing any loss of vegetation cover or any soil erosion), then the grass-fed beef produced in climates with year-round pastures would need just herding, providing water and (possibly) fencing, and no demand for the intensive cultivation of feed grains. In colder climates, the harvesting and storage of hay for winter feeding is also necessary.

But even grass-fed beef remains environmentally demanding, because ruminants require at least three times more plant mass per unit of weight gain when grazing than when they are fed grains—even when the comparison is done in terms of dry matter rather than as fresh weight—and at least 20–25 times greater mass of fresh grass or hay than pigs and chickens raised in CAFOs. When compared in terms of dry matter, grass and leaves eaten by ruminant grazers made the largest contribution to the world's 6.5 billion tons of livestock feed rations in 2020: nearly 50 percent. Crop residues (mostly grain straw and stalks) come next at nearly 20 percent, and feed grains contribute about 13 percent. This means that in 2020 about a third of global cereal production, and two-thirds of the US grain harvest, was fed to animals.[36]

Given the large amount of fresh (or dry) phytomass required to produce beef solely by grazing, it is not surprising that grass-fed beef has the highest water intensity of all major meats: the global average is 22,000 tons per ton of live weight.[37] This compares to 1,500–2,000 t/t for cereal grains and is of the same order of magnitude as producing the most water-intensive crop—coffee (about 18,000 t/t; see the third chapter). But the global production of coffee has been only about 10 million tons in recent years, while the worldwide production of beef has been more than six times larger, and most of these animals have spent variable parts of their lives grazing.

That is true even in the US and Canada, countries with by far the largest number of truly massive cattle CAFOs. Calves (born after artificial insemination to two-year-old cows after nine months of gestation) are raised with cows on pastures (without or with minimum grain) for six to ten months, then sold to cattle feeders when they weigh 270–360 kilograms. They are fed high-energy rations of grain and forage until they reach a market weight of 630–680 kilograms: this takes usually five months but it can be more than seven months. Heifers (young females before their first pregnancy) and steers (castrated bulls) can be left on pasture for anywhere between two and five years before reaching their slaughter weight.[38] Even if feed conversion rates in cattle were comparable to those in pigs or chicken, these long production spans would guarantee comparatively inferior performance.

The meat of the matter

Moreover, a proper management of grass-fed beef (above all, the prevention of overgrazing) requires, even in highly productive ecosystems, plenty of space. In North America the grazing areas required for a beef cow (their extent depends on the seasonal production of forage) range from just 1.1 hectares in green and rainy New England to 5.5 hectares on the northern Great Plains and 22 hectares in arid western mountain states, with the national mean about 7 hectares per animal; while in tropical Brazil, the common stocking rates are between 0.3 and 2 hectares per head. Even completely grass-fed beef puts a much greater burden on resources than pork or chicken.

Cattle's ruminant advantages cannot outweigh the burdens of their large size: slow reproduction and the long time to reach maturity (single births dominate; the frequency of twin-calving is less than 5 percent), and large aggregate basal metabolic needs. Clearly, cattle are not an optimal choice for a meat animal, and once the animals were not needed for draft the most rational decision would have been to keep them for milk but not for meat, except in those naturally grassy and rainy regions where they could be raised solely by grazing and without putting stress on water resources.[39] But after millennia of beef-eating by many of the Old World's cultures, and given the meat's popularity in 19th-century North America, that was never a realistic option.

Cows and climate change

But now an even more formidable charge against cattle has been added to the space, feed, and water claims: that of being a major global source of methane (CH_4), a greenhouse gas that absorbs outgoing radiation far more effectively (about 28 times more) than CO_2. Actual measurements of CH_4 emissions from Holstein cows in California (using airtight head and neck seals inside a large, clear plastic chamber) indicate that every year a cow emits 98 kilograms of methane.[40] But because of the ranges of animal sizes, ages, and specific diets, all global estimates of ruminant belching are just approximations. Given the large sizes of other US methane sources (mainly from fossil fuel combustion), American ruminants account for only about 2 percent of all direct emissions—and twice as much when indirect emissions, mostly nitrous oxide (N_2O) from fertilized soils, are included. The worldwide total remains uncertain.

The FAO's global assessment concluded that livestock supply chains emitted methane whose warming effect was equivalent to 8.1 billion tons of CO_2 in 2010, and that would imply about 8.7

billion tons in 2020.[41] Cattle contribute about 60 percent of the total, and (as expected, given large cattle numbers) Latin America is the largest regional contributor, followed by South Asia. When compared in terms of emissions per unit of protein, buffalo meat comes first, followed by beef. Relative emissions from milk production are only about 30 percent of those for beef, while pork, chicken meat, and chicken eggs are the least greenhouse gas–intensive sources of perfect protein.

Animal scientists have been trying to find out how to reduce methane emissions from beef and dairy production, and one promising option is the addition of seaweed to cattle's diet—but expanding seaweed harvesting to provide a daily feed supplement for the world's 1.5 billion cattle would be an unprecedented challenge. A much simpler way to reduce the volumes of water and methane associated with beef production would be to limit beef's consumption to meat produced by truly sustainable grazing in regions with high precipitation, and to divert grains now used in feedlots to produce more pork and chicken meat and more eggs instead. One key hurdle remains, however: people like eating beef.

Other options: water farming

Another option to reduce meat's impact on the terrestrial environment has been gaining ground: the expansion of aquaculture, or large-scale commercial fish production. Traditionally, this method of producing high-quality protein was largely restricted to herbivorous freshwater fish—above all, various species of carp in Asia, and also in parts of Europe—with saltwater aquaculture limited to shallow coastal seas (Hawai'i fishponds built with lava rocks), and to the production of crustaceans, mollusks, and algae in parts of East Asia. After half a century of post-1970 expansion, the aquaculture harvest of just over 85 million tons of fish, crustaceans, and mollusks has surpassed the capture of wild species in marine waters (just over

80 million tons), and it has been coming steadily closer to the total (salt- and freshwater) capture of 92 million tons.[42]

Herbivorous fish dominate the global aquaculture output, with China's four carp species (grass, silver, black, bighead) being a third of the total live weight production. The Nile tilapia (now mass-produced in Asia) is another important herbivorous species, as are various crustaceans and mollusks. In contrast to domesticated warm-blooded terrestrial animals, conversion efficiencies of feed to live weight are generally much lower in cold-blooded fish with significantly lower basal metabolic rates. For carp, the feed conversion ratios are between 1.5 and 2, for catfish between 1.2 and 2.2, and for tilapia between 1.4 and 2.4.[43]

This means that the lowest-rated are lower than the feeding efficiency for chicken (in terms of live weight), while the rate for edible portions will depend on dietary and cultural preferences—in Asia, carp may be eaten whole except for its bones. Diadromous species (salmonids, moving naturally between salt and fresh water, such as salmon and trout) account for two-thirds of marine finfish aquaculture, and feed conversion rates for these carnivorous fish are, not surprisingly, as low or even lower than those for herbivorous species—because protein and fats are easier to digest than cellulosic phytomass.

But the numerical equivalence of feeding ratios hides the obvious trophic level difference: herbivorous fish are, much like the poorest peasants in traditional societies, at the trophic level 2.0, salmon feed at 3.0, and tuna goes up to 4.0. While herbivorous carp can be fed inexpensive pellets prepared from mixtures of cereal and leguminous grains, carnivorous species will not grow rapidly and will not mature without being fed essential amounts of fish protein and fish oils, and these can be obtained only by catching smaller, less valuable herbivorous species (most often sardines, anchovies, and mackerel) and using them to make suitable fish feed. According to the FAO, in 2019 about 11 percent of the global fish harvest (some 20 million tons) was used for non-food products—above all, to make fish meal and fish oil for carnivorous aquacultured species. The expansion of fish-based feed has led to concerns about the overall efficiency of this enterprise, measured as the fish-in/fish-out (FIFO) ratio.

This ratio is 2.2 for marine fish, 3.4 for trout, and as high as 4.9 for salmon—meaning nearly 5 kilograms of wild fish would have to be caught to produce a single kilogram of salmon.[44] But separate consideration of fish meal and fish oil in the previous calculations does not account for fish meal left over after the production of fish oil. A corrected account leaves us with salmon fish-in/fish-out at 2.3, and trout, as well as the average for all marine fish, at 2.0.[45] This would still mean that the production of aquacultured carnivorous fish would entail a considerable loss of animal protein.

But as aquafeed production increased, companies changed the formulations, using more fish processing waste rather than freshly caught fish, and replacing significant shares of fish oil by plant oils. As a result, the total mass of globally produced fish meal and fish oil (some 6 million tons) remained relatively steady between 2000 and 2020, and there has been no additional pressure on marine resources to produce feed.[46]

Perhaps the best way to quantify the feeding burden of aquacultured omnivorous and carnivorous fish is to look at the ratio suggested in 2020 by a group of authors led by Björn Kok: the economic fish-in/fish-out ratio (also known as the economic FCR ratio) simply accounts for the amount of fish used to produce a kilogram of farmed fish.[47]

Advances in feed formulation and in aquacultural practices have steadily reduced this ratio for all species. For marine fish, eels, and carnivorous freshwater fish, the ratio declined from as much as 5.6 in 1995 to as little as 0.9 by 2015; for crustaceans, it fell from 2.6 to 0.5; and for salmonids (whose production is based completely on artificial feeding), it decreased from 3.8 in 1995 to 1.0 in 2020. This means that, by 2020, most aquacultured species (raised partially or solely by using fish-based feeds) were net producers of fish; eel was the only net consumer, while salmon and trout were net-neutral as their farmed production entails a 1:1 exchange of smaller feed fish for preferred oily red meat. And in 2017 it was reported for the first time that farmed salmon was producing more protein than it was consuming in wild-caught feed.[48]

But bluefin tuna—globally the most sought-after fish in the modern sushi-driven trade, and on the Red List of the most threatened species—is a different story. After decades of expensive trials and failures, it too can be raised from eggs spawned by tuna that were artificially hatched (so-called closed-cycle aquaculture).[49] But the practice, so far limited to Japan and one US location, is extremely demanding. Fewer than 1 percent of hatched eggs survive, and only 0.1 percent of hatched fish make it to harvestable age.

Big tuna eat about 5 percent of their body weight in fish every day, and hence the average FIFO ratio is very high: 10–15 for juveniles, and 20–30 during the years required for the fattening of adult tuna. A much more widespread form of tuna aquaculture is something that might be best called "wild tuna ranching": capturing small fish and then feeding them in pens (for up to 30 months) before their bodies reach the fat content preferred by the sushi market.[50] These pens are now operating in Japan, Australia, and several Mediterranean countries (Italy is the leader). But the practice does nothing for wild stock preservation as the young tuna are captured before they can breed, and it has the same, exceedingly high, FIFO ratio as the closed-cycle tuna aquaculture.

All kinds of aquaculture, from the shrimp ponds of Asia to the salmon ocean pens off the coast of New Zealand, have been accompanied (much like the animals feeding in terrestrial confinement) by concerns about the undesirable environmental consequences of raising fish in densely stocked pens, and of escaped non-native individuals breeding with local wild species. A radical answer to such concerns has been to produce genetically modified, rapidly growing salmon in containers on land. But it remains to be seen how far this route will take us.[51] Numbers of aquacultured marine fish species have been increasing, but fish popular in Asia still dominate: salmon currently ranks only 10th in the order of global annual production; and cod, another highly popular Western choice, is yet to make it (after some past setbacks) commercially.[52]

Should we eat animals?

One way to deal with this often-emotional topic is to refer the reader to a multitude of pro and con publications—although, as the contrast between two recent quotes demonstrates, that may just reinforce your preference. Nick Zangwill, a philosophy professor at University College London, states firmly that:

> If you care about animals, you should eat them. It is not just that you *may* do so, but you *should* do so. In fact, you owe it to animals to eat them. It is your duty. Why? Because eating animals benefits them and has benefitted them for a long time. Breeding and eating animals is a very long-standing cultural institution that is a mutually beneficial relationship between human beings and animals.[53]

Gary Francione, a professor of law at Rutgers University in the US, is no less firm in the very opposite fashion, and he extends the proscription to all animal foods:

> And it's not just meat that is a problem; there is no morally significant difference between meat on the one hand, and dairy and eggs on the other. *All* of these products involve suffering and death. Veganism is not an extreme position; what is extreme is claiming to believe that animals matter morally and then inflicting suffering on them for no reason other than culinary pleasure or convenience.[54]

All of these products? Shepherds who take care of free-grazing flocks, milk them, and make traditional artisanal cheese might have something to say about their charges being exposed to suffering and death.

Where do I stand? I am always mindful of Theodosius Dobzhansky's fundamental dictum: "Nothing in biology makes sense except in the light of evolution,"[55] and it is an incontrovertible evolutionary and physiological reality that we have descended from a long line of omnivorous primates and hominins, that our digestive system is indubitably that of an omnivore, that our evolution and history during the past 11,000 years is closely linked with the domestication of more than a dozen mammalian and avian species, and that our health and mental development have benefitted from consuming their meat, eggs, and dairy products. That we should treat animals humanely (and still too often do not) and that we should consume animal foods in moderation are obvious caveats supported by a great deal of evidence—but evolutionary perspectives do not justify seeing the domestication of animals as regrettable or the eating of animal foods as despicable.

5. What's More Important: Food or Smartphones?

For decades, gross domestic product has been the leading measure of modern economic progress—politicians always want to see it going up by whatever they consider to be a "healthy" rate. As a result, even in large, already affluent economies where average annual per capita incomes are in the tens of thousands of dollars, anything below 1 percent is seen as disappointing and undesirable. And economists swoon over countries whose annual GDP growth approaches 10 percent; if it gets into double digits, they become outright giddy. No wonder that during the past three decades, they reserved their greatest admiration for Communist China, whose GDP growth was mostly between 9 and 14 percent between 1991 and 2010, before declining to the 6–7 percent range during the 2010s—very similar to India's economic growth.[1]

And what do these GDP numbers tell us about food production? That by modern economic standards it is nothing but the most marginal of all human endeavors; that when it is judged by the share of the total economic output it is the least important contributor to economic activities of all modern societies. Its share in the global economic product has been steadily decreasing: in 2020 it was just 4 percent (compared to about 10 percent in 1970), with relatively high shares (>10 percent) prevailing only in Africa and the poorest nations of Asia, while the contributions in the UK, Germany, the US, Japan, and France are, respectively, just 0.8, 0.9, 1.0. 1.0, and 1.9 percent.[2] Every sector is economically more important: construction, transportation, manufacturing, and, of course, the now ubiquitous "services," all of which account for 65 percent of the global economic product and contribute 77 percent of the total in the US. This means that in 2020, with the annual global economic product at about $85 trillion, worldwide services were valued at more than $55 trillion, while agriculture contributed less than $4 trillion.[3] But this ranking of economic importance does not make sense—and it is demonstrably wrong.

Smartphones vs. staple grains

That it does not make sense—that it is the result of valuations that do not distinguish between irreplaceable necessities and dispensable, even frivolous activities—can be easily demonstrated by comparing just a few numbers. The global market for financial services is now worth more than $20 trillion, nearly a quarter of the total economic product, and losing a fifth of it (the equivalent of the total global agricultural output!) would put its size where it was about a decade ago. That would cause some financial dislocations—but unlike losing a fifth of the global food output (that is food for more than 1.5 billion people!), the change would not result in famine or mass mortality. Another way to highlight the absurdity of our economic valuations is to compare the value of smartphones and staple grains.

The global smartphone market was worth about $400 billion in 2021—and that was only about 10 percent less than the value of the global wheat and rice harvest in the same year (calculated by assuming average annual prices of, respectively, $260 and $460 per ton).[4] The sudden disappearance of mobile phones would, obviously, cause some problems, but the required adjustment (after all, according to this thought experiment, the internet would remain intact) would be incomparably easier than the loss of 1.3 billion tons of the two most important staple grains, which would result in unprecedented famines and the deaths of a significant portion of today's 8 billion people.

History vs. the economists

In order to show how the current valuation of what the economists count as "agriculture" is flawed, it is necessary to deconstruct and reconstruct the sector's definition and extent. The word's Latin origin—*agri cultura* (*ager* being a farm or simply "land")—restricts it to crop cultivation, but long ago it became obvious that domesticated animals (providing draft, food, and fertilizer) are an integral part of the endeavor. Moreover, modern economic definitions of the sector also add harvesting fish (from managed or natural habitats), and in some national statistics the value of forestry is added as well, because economists see it as a kindred "extractive" activity. To anybody even remotely familiar with the history of economic activities, such narrow definitions would be good only for traditional agricultures where more than 90 percent of all people live in villages or small towns and are engaged in planting, tending, harvesting, processing, and selling crops, and taking care of domestic animals and earning income by selling meat and dairy products.

Men, women, and children (as young as four years old) worked in the fields—sowing, weeding, manuring, and harvesting fruit by hand, reaping crops with sickles or scythes, carrying sheaves on their backs or in animal-drawn carts, and tending animals (stabling, herding, and feeding). Then they worked on threshing grounds and in their yards, stables, and homes: processing the harvests, hand-milling grains, extracting edible oils in presses, milking cattle, goats, sheep, yaks, or camels, making butter, cream, and cheese (all by hand), drying fruit, killing animals, preparing longer-lasting meat products, or simply smoking and drying meat. But even those ancient agricultures relied on inputs not made by farmers, herders, or fishers themselves. They did not, for example, dig up iron ore and smelt it with charcoal in simple, small shaft furnaces to make the metal into knives, sickles, scythes, hooks, or hoops—those inputs came from artisanal manufactures.[5]

The variety of external inputs (those not made by food

producers themselves) increased with the deployment of draft animals (which required harnesses capable of providing effective power for field work and road transport), with the larger-scale processing of seeds (grain-milling and oil extraction powered by water mills and later also by windmills), and with the growing trade in foodstuffs. Rare are the discoveries of Roman shipwrecks that do not contain plenty of amphoras, the famous two-handled pottery storage jars used to transport olive oil; the mass-scale manufacture of these containers required the mixing of select ingredients and proper kiln-firing.[6] And as noted in the second chapter, imperial China went to extensive efforts to build and maintain a country-wide network of large grain stores.

The growth of settlements led to the emergence of intensive peri-urban cultivation of vegetables (fertilized with urban wastes, an arrangement that survived around many cities until the 20th century, before being displaced by imports from distant places) and the increased adoption of out-of-home preparation and eating of foods. Baking bread (often in large commercial bakeries) and preparing ready-to-eat cold or warm meals (in Greek and Roman, *thermopolia*) required plenty of labor to gather fuelwood (or to fire charcoal) and to deliver it to cities and towns, and skilled work to build ovens and stoves and to bake, cook, sell, and serve the food.[7]

This extension of inputs that originate outside farming (or herding or fishing) but are indispensably connected to producing, trading, preparing, and consuming food both intensified and spread to more countries in the early modern era (1500–1800), with the emergence of a truly world-spanning trade in spices, tea, cocoa, and sugar; rose to new heights during the decades of rapid industrialization during the late 19th and early 20th centuries; and reached unprecedented intensity and a truly global extent after the end of the Second World War. Any inhabitant of a larger city in any high-income country does not have to go to specialty food stores to buy fresh produce, grains, cheeses, condiments, meats, and prepared foods that originated in scores of countries, from Peruvian oranges to Turkish dried apricots and from Greek feta to New Zealand lamb. Given these increased imports, it is not surprising that during the first two decades of the 21st century, the global volume of agricultural exports nearly doubled, and their value tripled.[8]

The 1 percent

These realities make it clear that standard economic accounts claiming that agriculture adds less than 1 percent to the total value of national GDPs are very poor indicators of the real value and the real material and energetic extent of modern food systems, of their importance to modern economies—and of their environmental impact. At the same time, there has been no binding, internationally acceptable agreement as to a single, clear definition of the real extent of food-producing and food-consuming activities. While there is no consensus about how far one should expand those more realistic accounting boundaries when studying national food provision or the global production/consumption system, the minimum extension should consider the contributions of all direct and indirect inputs of energy that have become indispensable to the production of food in modern societies.

This group of products and services has been steadily expanding, and for food production it must now include the following energies, products, and services.

Field operations

This encompasses the production and distribution of farm machinery, ranging from tractors, combines, and trucks to the many specialized seeding, cultivation, and harvesting implements, and the irrigation pumps and center-pivot systems and fuels (mostly diesel oil, also gasoline and liquefied petroleum gas) to operate it.[9] In non-irrigated crops, it is usually the synthesis of nitrogenous fertilizers (and preparation of phosphates and potassium applications) that embodies most of the indirect energy required for cropping.

Pesticides and herbicides and the development and production of seeds (for modern high-yielding cultivars they must be bought anew every year, not saved from previous crops) require much smaller energy inputs. Finally, there are both direct and indirect energy demands for the on-farm storage of harvests prior to sale, including structures (bins, silos) and equipment (dryers, elevators) and the electricity or diesel oil required to operate it.

Livestock production

This includes the preparation of commercial feeds, additives, and veterinary drugs, the construction and operation of stables, barns, grow-out houses for birds, and feedlots, and the energy needed for heating and cooling these enclosed spaces and for operating mechanized waste management and water treatment.[10] For livestock raised on pasture, it is the cost of emplacing and repairing grazing enclosures and providing water and supplementary feed (in times of drought or during severe winters).

Fisheries

For aquatic products we would have to add the cost of fishing ships and their gear (nets, lines), port and repair facilities, and the cost of diesel fuel (now almost all marine engines run on diesel). For aquaculture, it would be the cost of building, maintaining, and operating (aerating, cleaning, repairing, dredging) ponds, pens, and cages, breeding species from eggs, and feeding them until they reach the market weight.

Machines are more important than smartphones

How much these sectors have become diversified can be shown by taking a closer look at the US agricultural machinery market, which now includes tractors, plowing and cultivating implements (plows, harrows, cultivators, tillers), machines for planting (seed drills, planters, spreaders) and harvesting (combines, forage harvesters, and specialized equipment used to harvest vegetables and now even grapes), haying and forage (mowers, balers, tedders, rakes), irrigation (sprinklers, center pivots, drip irrigation), and other machinery used for ditching, dike construction or field leveling. Given the extended heavy-duty use, agricultural machinery requires proper maintenance and often significant repairs, resulting in additional material and energy inputs.

Available accounts of these outlays differ. In 2020 the market size (measured by revenue) of America's tractor and agricultural machinery industry was about $40 billion, while the summary of farm production expenditures shows that in the same year American farmers paid about $25 billion for tractors, trucks, and other machinery, as well as nearly $20 billion for other farm supplies and repairs.[11] The other major direct material and energy inputs into food production were $26 billion for livestock, poultry, and their feed, nearly $25 billion for fertilizer (adjusted for net trade), $23 billion for seed and plants, nearly $17 billion for agrochemical preparations (herbicides, insecticides, fungicides), and $11 billion for fuel.[12] Obviously, these inputs claimed significant shares of the nation's output in several major industrial sectors (above all in the machinery and chemical industries).

The price of production

Food has an even greater economic impact after its production. Its processing, packaging, storage, trade, transportation, wholesale, retail, preparation, and waste disposal reach deep into every sector and multiply its overall economic importance. A recent study of the agriculture and food sector in the US economy performed a more inclusive valuation, as it considered all sectors whose added value is based on crop and animal production: all food and beverage manufacturing, food and beverage stores, food service and eating and drinking places, as well as textiles, apparel, and leather products, and forestry and fishing.

According to this definition, the agriculture, food, and related industries added $1.055 trillion to the US GDP in 2020, or 5 percent of the total, and they employed 19.7 million people (part- and full-time) or 10.3 percent of the country's labor force, with food service and eating and drinking places accounting for slightly more than half (5.5 percent) of that share.[13] This account raises the importance of the agriculture and food sectors from the undervalued 1 percent share to a more realistic 5 percent share in terms of annually added value, and to a 10 percent share in terms of total employment—but the actual food-related share in that account is smaller because this study also includes a number of clearly (or largely) non-food enterprises, including tobacco, leather, textiles, and forestry.

The differences between agriculture's shares of GDP and its shares of employment are far larger for low- and medium-income countries. In 2020 China's agricultural sector still employed nearly 24 percent of the country's labor force, and official data show an additional 2.5 million people in agriculture and sideline food processing, 1.6 million in food manufacturing, and 1.1 million in the production of beverages.[14]

The cost of eating

Another option to gauge the impact of food provision on everyday life is to look at food expenditure as a share of an average household's disposable income. This share is subject to a universal rule clearly formulated in 1857 by Ernst Engel, a German economist and statistician: "The poorer the family, the higher its share of expenditures spent on securing its nutrition."[15] This rule is valid within societies—the ratio drops as people become richer—as well as among nations (the poorer ones spend more than the more affluent countries). The US share declined from about 43 percent in 1900 to just 8.6 percent in 2020 (but by 2022 it had risen above 11 percent);

the Chinese share remains considerably higher (almost exactly 30 percent in 2020) but is now less than half of where it stood before the beginning of economic reforms in 1980. The Indian rate is very close to the Chinese mean, and within the EU the ratios follow a predictable ranking, from Romania's 26 percent to Germany's 10.8 percent.[16]

A waste of energy

But even that total is still too low, and perhaps the best (although inherently rather difficult) way to get closer to the real share is to assess the fundamental importance of the entire food system in a national—or global—economy, by establishing its aggregate energy cost. Energy valuations provide a more fundamental measure of physical importance than monetary assessments do, and can be compared without any of the biases accompanying currency conversions and valuations.

Calculating all direct and indirect energy inputs is even more daunting than assembling relevant financial accounts, however, and in the absence of any clear rules, the outcome depends, again, on setting specific analytical boundaries.[17] Remarkably, detailed US energy balances trace the consumption of all energy sources for residential, commercial, industrial, and transportation categories, but do not offer any estimates for either narrowly defined agricultural or broadly defined food-related uses. The International Energy Agency singles out agriculture (combined with forestry) in its annual national balance accounts, but it shows only direct inputs (predictably dominated by liquid fuels for farm machinery) amounting to very small shares of the national primary energy supply. The approximate values range from 0.1 percent in Japan to 1 percent in the US, 1.7 percent in France, and 2.3 percent in Canada, with the global mean coinciding with the Chinese share of 1.4 percent.[18]

There is a considerable difference when including only direct energy inputs (fuels and electricity for field machinery and irrigation pumps, and for the heating and air conditioning of structures housing animals) and when also quantifying major indirect energy needs (fuels and electricity used during the production of already listed indispensable inputs, ranging from machines and fertilizers to agrochemicals and irrigation pumps). A recent study of direct and indirect energy uses in US agriculture showed that, despite expected annual fluctuations, there has been a fairly steady demand for about 1 exajoule (10^{18} joules, equivalent to 25 million tons of crude oil) a year of direct energy (diesel, gasoline, natural gas, liquid petroleum gas, electricity), and the equivalent of about 18 million tons of oil for indirect uses (such as for the synthesis of agrochemicals and for lubricants).[19]

Combined, this would be less than 2 percent of America's high energy consumption, which is dominated by the transportation and industrial sectors. But the study leaves out several important indirect energy inputs, such as the embodied cost of steel, aluminum, plastics, and glass—materials that are required to make agricultural machinery—and its boundary ends at the farm gate. Once again, that cut-off was fine for traditional agricultures, where all but small shares of crops and animals were consumed by subsistence peasants and less than 10 percent of the entire production was claimed by people living in towns and cities—but it is not fine for modern food production

The most realistic, most comprehensive analytical boundary designed to capture the complete universe of food-related energy consumption would embrace all direct and indirect activities connected with production (of plants, animals, and fish), processing (from milling to canning), trade (all forms of intra- and international transfers, from railways to jetliners), storage (from short-term to seasonal to long-term refrigeration), distribution (now mostly by trucking), sales (from farmers markets to grocery chains), preparation (at home, in institutions and restaurants of all kinds and sizes), consumption (at home, eating out), and post-consumption measures (garbage disposal, waste treatment).

The best data we have

In 2010, the US Department of Agriculture published a study whose analytical boundaries came close to this system, looking at crop and animal production, at the processing, packaging, transportation, wholesale, and retail of food, and at food services and household food purchases and preparation, as well as food-related waste disposal. Household operations had the highest food-related energy use, while energy used for food processing showed the largest rate of growth.[20] The study traced all energy uses in the United States in three interrelated steps: measuring all energy directly used in domestic production activities, including household operations, according to roughly 400 industry classifications; tracing energy flows embodied in all energy-using products throughout the economy, ending with final market sales; and identifying all food-related markets and assessing the food-related energies embodied in final sales. The study concluded that food-related energy use as a share of the country's total primary energy consumption grew from 14.4 percent in 2002 to an estimated 15.7 percent in 2007.

What adjustment should we make for 2019, the last year not affected by pandemic lockdowns and economic downturn? In 2019 the total US energy use was about 2 percent lower than in 2007, while spending on food had risen by nearly half, with food-at-home spending reaching $808 billion (and $876 billion during the first pandemic year) and food-away-from-home spending (purchased from restaurants, fast-food places, schools, and other away-from-home eating places) approaching $1 trillion ($978.2 billion).[21] And as this rise took place during a period of low inflation, nearly all of that increase was real.

The harvested area and the use of fertilizers in the US have hardly changed (differences on the order of a mere 1 percent), but food imports—driven by horticultural products, from coffee and fruits to nuts and wine—rose substantially (also including more wintertime shipments of highly perishable fruits and vegetables from warmer climates), as did the energy demands for packaging. These changes would have resulted in a slight increase of the share, perhaps up to 17 percent of the country's primary energy use.

But the most important change has been the decision taken by the US Environmental Protection Agency to estimate more comprehensibly flows of food throughout the country's food system by expanding the accounts beyond composting, combustion with energy recovery, and landfilling.[22] The new methodology, used for the first time in 2018, includes food waste going into animal feed, biochemical processing, land application, anaerobic digestion, and sewer/wastewater treatment. As a result, in 2018 food waste accounted for nearly 22 percent of the total municipal waste stream, compared to just 13 percent in 2005—adding another percentage to the food system's overall energy claim, putting it as high as 18 percent, and so the range of 15–20 percent seems to be a highly defensible conclusion.

Beyond the US

In China, the agriculture (and forestry) sector has recently consumed directly only about 2 percent of all primary energy, and fragmentary information on direct food-related household, food services, and transportation uses raises the food system's share of total energy use to about 12 percent; the addition of energy used in food processing and the production of beverages raises the total to 14 percent, and the production of fertilizers adds another 2 percent.[23] Even conservative assumptions regarding energy claimed by food and feed transportation (China is now a major importer of animal feed) and by food storage (the country keeps the world's largest government-supervised stores of grain and edible oils) raise the grand total to 20 percent of the country's large (now the world's largest) primary energy supply—very similar to the US fraction.

In the absence of relevant data for most of the world's countries, the analogical global share can only be a best possible estimate with considerable margins of uncertainty. In many low-income countries, shares of direct and indirect energy use in agriculture will be relatively high because much larger shares of the population are still employed in agriculture and in fisheries, intensive fertilization is needed to feed large populations, and private energy consumption for housing and transportation remains comparatively low. Most families in low-income countries will also spend relatively more energy on cooking (with too many families still using highly inefficient stoves burning wood, charcoal, or straw) but they will eat food subject to much less food processing, packaging, and long-distance distribution.

My estimate of global direct and indirect use in the production of plant and animal foods (including fishing) was about 17 exajoules (EJ)—equivalent to some 400 million tons of crude oil—at the beginning of the 21st century.[24] In 2011, an FAO study, relying heavily on my calculations, put the energy cost of global food production at about 20 EJ, and added about 40 EJ for food processing and distribution and 35 EJ for retail, food preparation, and cooking: the total of 95 EJ (equivalent to 2.2 billion tons of crude oil) was about 30 percent of the global primary energy supply.[25] With the intervening rise in agricultural output and an improvement in various conversion efficiencies, I would raise my estimate of energy use (direct and indirect) in crop and animal production by about 25 percent for 2020, and then add at least an additional 10 percent of that total in order to account for energies consumed in fisheries and aquaculture: that would bring the total close to 25 EJ. An identical amount of energy used by farming and fishing is needed to process harvested crops (dominated by cereal milling, oil pressing, and sugar refining) and animal foods (dominated by butchering and dairy production).

The wheels on our meals

There are no comprehensive global statistics on transportation volumes by commodity. In the US, food and beverages account for about 18 percent of all annually transported mass, and additional fertilizers and agricultural machinery bring that to about 20 percent. A conservative global estimate can be derived as follows: taking the global harvest of all major crop varieties (cereals, legumes, tubers, oilseeds, sugar crops, fruits, and vegetables), a total of about 9.3 billion tons in 2020; subtracting the international exports of about 800 million tons; for the domestically consumed food assuming an average distance of just 1,000 kilometers, before and after processing and storage, and multiplying that by a typical energy cost of rail and road freight; then assuming an average export distance of 6,000 kilometers for the exported food and multiplying that by an average energy cost of shipping and rail freight (shipping has much lower costs than rail).[26] This adds up to at least 25 EJ of energy—that is as much as the energy cost of cropping and fishing, and the equivalent of nearly 5 percent of global energy consumption in 2020.

Costs of food preparation

The last major item to be added is the energy cost of food preparation and refrigeration, in households and by food services. Given the enormous disparities in the ways food is prepared between affluent and low-income countries—and their different efficiencies—it is difficult to offer an average multiplier. Studies in the EU show that cooking totals nearly 6 percent of household energy consumption, and that amounts to about 27 percent of the EU's total energy use.[27] As expected, the US per capita mean is somewhat higher, and the average rates in China and India are higher still because many rural families rely, as elsewhere in low-income countries, on the inefficient combustion of woods, straw, and charcoal: in urban China, around 50 percent above the US mean; in rural China, about 2.5 times as much, with similar rates indicated by Indian studies.[28]

Conservative assumptions of average per capita requirements for cooking would result in the global annual range of 21–28 EJ.[29] Nearly 2 billion refrigerators and freezers are now storing food around the world, and when assuming an annual electricity consumption of about 350 kWh/year per refrigerator they would require at least 2.5 EJ of electricity.[30] That would translate to more than 5 EJ of primary energy use—and it would raise the global cooking and refrigeration demand to 26–33 EJ a year. The addition of industrial-scale refrigeration (required for the prolonged storage of meat, fish, and butter, and for intercontinental shipments of these commodities) could bring the upper range to 35 EJ and the global food-related energy demand to no less than about 20 percent (115 EJ) of the global primary energy use in 2020.

But this global account is still missing such indirect energy costs as the production of agricultural machinery (tractors, combines, implements, irrigation systems and trucks made of steel, aluminum, plastics, rubber), fishing vessels and aquacultural ponds and pens, seed breeding and cultivation and the cost of extension services (now an indispensable component in crop cultivation and animal husbandry in all major food-producing countries), as well as the cost of waste disposal. As a result, it is safe to conclude almost certainly that even a very conservatively assessed energy cost of the global food system is no less than 20 percent and most likely on the order of 25 percent of the world's recent annual primary energy supply.

Something's not adding up

Although the food system's complexity and the paucity of data make it impossible to calculate accurate shares of aggregate economic production, total expenditures, or energy requirements that can be attributed to global production, processing, transportation, wholesale, retail, storage, and consumption of food, there is no doubt that the actual values are all on the order of 25–30 percent of respective totals, and that the standard economic accounts of agriculture and fishing contributing just 1–4 percent to the value of the global economic product are among the best examples of grossly inaccurate and highly misleading quantifications.

Environmental impacts

The standard economic accounts attributing less than 5 percent of the world's annual economic product to food production also provide an unrealistic perspective as far as the environmental impacts of the global food system are concerned: all of them claim far higher shares. Agriculture has a dominant claim on the planet's water resources: cropping and animal production require 72 percent of the world's water (surface and underground) withdrawals.[31] Food production is also the world's largest category of land use, with the land under annual and permanent crops now surpassing 1.5 billion hectares, and grazing land (pastures) claiming an area more than twice as large (3.1 billion hectares) for a total equivalent of about 36 percent of non-glaciated land.[32]

The increasing use of nitrogenous fertilizers amounts to the largest human interference in the complex global nitrogen cycle. In the early 2020s, about 110 million tons of nitrogen was applied annually to crops, in compounds derived from the Haber-Bosch synthesis of ammonia, compared to no more than 40 million tons of nitrogen left in the soil by nitrogen-fixing leguminous crops and free-living bacteria, and about 25 million tons of nitrogen in recycled organic wastes (mostly animal manures).[33] Nitrogen losses (above all through the volatilization of ammonia, water runoff, and the leaching and erosion of nitrates) are also a major contributor to the acidification of fresh waters and to the creation of dead zones in some coastal areas that receive runoff from heavily fertilized fields.[34]

And the food system makes a large contribution to the generation of greenhouse gases. Given the recent concerns about the progress of global warming, there are many national accounts and global estimates of greenhouse gas emissions originating in crop and animal production, as well as those considering the contributions of the entire food system.[35] By far the most comprehensive recent study, with data assembled for 2015, considered the entire global food system (from production to consumption, including processing, transport, and packaging), and put its total greenhouse gas emissions at 18 billion tons (Gt) of CO_2 equivalent (uncertainty range of 14–22 Gt) or 34 percent (uncertainty range of 25–42 percent) of total emissions in that year.[36]

Almost 40 percent of the total came from agricultural inputs (dominated by fertilizers); nearly a third of these emissions originated in land use and land use changes (deforestation to expand farmland and grazing land, degradation of organic soils); and 29 percent were traced to transport, processing, packaging, retail, consumption, and waste disposal. As for the gases, CO_2 dominates (52 percent), with methane (CH_4) contributing about 35 percent of all emissions (compared in terms of CO_2 equivalent), and nitrous oxide N_2O (mostly from the denitrification of nitrogenous fertilizers) about 10 percent. Estimates of global methane emissions from livestock are particularly difficult to make, but that has not prevented many accounts portraying cows as the greatest threat to humanity's survival.[37]

At what cost?

Trying to express these burdens in monetary terms is, inevitably, an exercise ruled by approximation and by (more or less defensible) assumptions. In 2019, the World Bank's Martien Van Nieuwkoop tried to compare the food system's monetary value (he put it at about 10 percent of the global economic product) with its costs: he considered the health and social costs due to some 2 billion undernourished (lacking energy or specific nutrients) people, estimated at 3 percent of the world's economic product; the costs of obesity (2 percent of the global economic product) due to disease and premature death; and of wasted food production, inadequate food safety, lost or damaged terrestrial ecosystems (including land degradation), and food-related greenhouse gas emissions. This was summed up to $6 trillion in 2018, equivalent to more than 7 percent of that year's global economic product and "a bill that is simply too high for $8 trillion worth of food."[38] How much higher, or lower, is the actual value? Supporting arguments can be found for either conclusion.

The rising cost of obesity is a food-associated burden that has received plenty of attention in recent decades. The most up-to-date estimate of the annual costs of obesity is 3.3 percent of GDP as the Organisation for Economic Co-operation and Development (OECD) average, with values as high as 5 percent in Brazil and Mexico.[39] Realistic valuations of lost or degraded ecosystems and their services (ranging from biodiversity that promotes long-term stability to pollination and water retention) have been notoriously elusive.[40] And the relative cost of reducing CO_2 emissions (and hence the consequences of their rise) ranges across three orders of magnitude, from the most affordable opportunities for tropical reforestation to the most demanding subsidies for photovoltaic electricity generation with requisite increases in grid capacity.[41]

In contrast, the most comprehensive attempt to quantify national external costs of agricultural production would support a lower value of the global burden. In 2004 two economists at the University of Iowa attempted to quantify these costs in terms of damages to water, soil, and air resources (including greenhouse gas emissions), and to wildlife and human health due to pathogens and pesticides.[42] As expected, their final cost estimates ranged widely, with the annual total in the US as low as $5.7 billion and as high as nearly $17 billion in 2002 dollars, and with the largest share attributable to damaged soil resources, ranging from the replacement of reservoir capacity lost due to soil erosion to increased flood damage. Moreover, at that time government payments to regulate agricultural production and to mitigate some of its damages amounted to nearly $4 billion a year.

But in 2002 US agriculture added at least $100 billion to the country's economic product (amounting to only about 1 percent of GDP), and hence even the highest damage ($17 billion) would be equal to less than a fifth of the annually added value. And when comparing this damage to agriculture's more realistic contribution to the country's economy (on the order of at least $1 trillion), its share would be less than 5 percent. And even if analogically defined damages in many low-income countries of Africa and Asia (where environmental deterioration has been more extensive) would be three or four times this enlarged US share, we would come nowhere near to Van Nieuwkoop's high cost/benefit ratio of 0.75 ($6 trillion vs. $8 trillion), as the environmental and health damages would add up to less than 20 percent of the food system's real value.

A sense of urgency

Attempts at quantifying the annual burden that the global food system imposes on the Earth's environment and on human health, and adding to this the value of investments designed to control, minimize, or eliminate such undesirable effects, will remain open to easily justifiable criticism—and it might not only be elusive but even counterproductive. We know enough about the functioning of the biosphere to be highly concerned about environmental deterioration caused directly and indirectly by food production, and we also know, without waiting for any new monetary estimates, that many of these changes require overdue action.

When the FAO's first report on *The State of the World's Land and Water Resources for Food and Agriculture* (SOLAW) appeared in 2011, its subtitle was neutral: *Managing Systems at Risk*.[43] The report's second edition, published ten years later, is subtitled in a highly concerning but certainly not exaggerated way: *Systems at breaking point*.[44] Three broad conclusions justify this concern.

In too many regions the interconnections between land, soil, and water—all indispensable to the production of food—are nearing their limit. Recent paths to agricultural intensification (including the increasing use of agrochemicals and irrigation) now give diminishing returns in many of the world's breadbaskets, and further increases are not sustainable for both environmental and economic reasons. At the global scale, food-producing systems have been excessively polarized: large commercial enterprises dominate agricultural land use, accompanied by the continued fragmentation of small subsistence farms on lands susceptible to soil degradation and water deficits.

Given the range of specific concerns, even a concise review would require another book. Instead, I have selected several degradative trends that illustrate the concerns that I see as particularly worrisome. They include the continued deforestation of the Amazon, the decline of India's groundwater resources, the contamination of China's agricultural soils, the excessive presence of reactive nitrogenous compounds in the Netherlands, and the increasing aridity of the US Southwest.

Deforestation

Few environmental problems have received such widespread media attention as the deforestation of the (mostly Brazilian) Amazon, driven by the quest for extending agriculture (above all soybean cultivation for export) and grazing for beef cattle: its annual rate peaked in 1995 at nearly 30,000 km^2 a year (almost the size of Belgium), and after a brief decline rose again to 27,800 km^2 by 2004.[45] The subsequent reduction was driven not only by new government policies but, more importantly, by the Amazon soy moratorium (a ban on the purchase of soybeans from newly deforested areas) and agreements in the cattle sector.[46]

This brought the annual rate down to just 4,500 km² in 2012 (an 84 percent cut from the previous peak), but, once again, this was followed by a new period of increased deforestation: its rate doubled between 2012 and 2019 and by 2021 had risen to about 13,000 km², equal to nearly two-thirds of Wales. Perhaps the most important long-term consequence of continued deforestation is the reduction of Amazon basin rainfall. The effect remains uncertain, but the most recent simulations suggest that the continuation of high (pre-2004) deforestation rates would result in an 8 (±1.4) percent greater reduction in the basin's mean annual precipitation than natural variability by 2050.[47]

Groundwater depletion

Groundwater depletion in India, now the world's most populous country, has been well studied and shows a large spatial variability. Satellite monitoring shows that since the mid-1990s the territory north of 25°N—including not only the more arid Northwest (the states of Rajasthan, Haryana, Punjab) but also the northeastern states of Arunachal Pradesh and Assam (the country's rainiest region)—has shown a significant decline of groundwater storage, amounting to about 15–25 cm/year, with withdrawals exceeding the groundwater recharge during normal monsoon years. In contrast, Madhya Pradesh, Maharashtra, and Andhra Pradesh show either little change or gains of up to 10–20 cm/year.[48] These conclusions have been supported by the monitoring of wells. A decreasing precipitation trend has been seen only in Tamil Nadu, Kerala, and Karnataka, and despite an overall increase in precipitation, the rapid depletion of groundwater storage has been recorded not only in Haryana but also now in Assam.[49]

Heavy metal contamination

Unlike the Amazonian deforestation or concerns about the depletion of aquifer (water-bearing rock, be it in India or on America's

Great Plains), the heavy metal contamination of China's soils has received comparatively little international attention. But the problem is extensive: by 2014, 20 percent of agricultural land and 10 percent of forested land in China was contaminated by one or more heavy metals, above all by cadmium (from phosphate fertilizers and the emissions from coal combustion) and by arsenic, mercury and lead.

Unfortunately, the provinces with the highest rates of soil contamination are also the leading producers of staple cereals. As a result, nearly 15 percent of China's grain production has become affected by heavy metal contamination, with Hunan province—producing 15 percent of China's rice but responsible for nearly three-fifths of all food contamination by mercury, a third of all cadmium, a quarter of all lead, and a fifth of all arsenic emissions—having the worst problem.[50]

Affluent countries

While affluent countries—many now with stagnant or even declining populations, surplus food production, and commensurately high rates of food waste—are in a better position to modify their agricultures in more rational and less destructive directions, they too must deal with some difficult challenges, be they caused by past malpractice or by environmental change. A good example of the first problem is the Dutch surfeit of nitrogen, while the increasing aridity of the southwestern US states presents a major difficulty for the world's largest food exporter.

The exceptional degree to which the Dutch are drowning in animal waste is easily illustrated by comparing the numbers of large animals per unit of all agricultural land (arable and pasture) or per unit of cropland. In 2020 the Netherlands had two cows per hectare of agricultural land and 11 pigs per hectare of cropland, while the analogical rates were 0.5 and 0.8 for the UK and 0.25 and 0.5 for the US.[51] When all animal (cattle, pigs, sheep, poultry) are converted to equivalent weight, the EU averages 0.8 animal units per hectare: Bulgaria has just 0.2, but the Netherlands has 3.8 animal units.[52] No matter how well managed, such densities produce too much nitrogen pollution per hectare, and on December 15, 2021 the Dutch coalition government decided to deal with the problem in a radical way—by reducing the country's livestock by a third.[53] The government allocated €25 billion for buyouts, the relocation of animals, or, as a last measure, the expropriation of livestock. Not surprisingly, the decision was widely opposed by farmers.

In the US Southwest, the dominant environmental concern is too little precipitation, insufficient snowpacks in the Rockies and Sierra Nevada, and extensive multi-year drought. By the end of March 2022, all of Nevada and Utah, nearly all of New Mexico, most of Oregon and western Texas, and all but a tiny corner of California were experiencing no less than severe—and mostly extreme and even exceptional—drought.[54] Reconstruction of soil moisture deficit, based on tree-ring observations and going back to 800 CE, suggested that this drought of 2000–2018 was exceeded only by a megadrought in the late 1500s.[55] The drought persisted in 2022 but was substantially reversed by record precipitation during the first three months of 2023.[56] Whatever the future might bring, any drought period has become more worrisome in the region that is now home to nearly 40 million people and that, in terms of value, is the country's leading agricultural producer.

This is perhaps the most concerning reality of these degradations: that the worst affected regions are either densely populated or are major food-producing areas (or both), where the demand for food and the intensity of cultivation are very high. Similar problems in other regions are negatively affected by political instability and recurrent conflicts, with the Middle East and most of Africa being the chronic examples of such problems. As I will note in this book's closing chapter, these undesirable trends are reversible, or at least sufficiently manageable, but not without substantial and persistent course adjustments.

The grand account cannot be very accurate, but there is no question that standard economic appraisals have massively undervalued both the contribution and the cost of the global food system—and in the next chapter we will see what the nutritionists have got wrong.

6. What Should You Eat to Be Healthy?

In this chapter I will attempt a version of the proverbial separation of grain from chaff, blowing away the chaff of dubious nutritional claims, miraculous diets, and supposedly life-changing dietary supplements, and pointing out the basic grains of human food requirements and the actual food intakes that provide the best foundation for a healthy life and admirable longevity. And I will examine how this understanding can be translated into long-term dietary recommendations.

This task is made easier thanks to the accumulated understanding of human food needs and by amassed statistical evidence supporting some clear conclusions, but these realities must reckon with a flood of claims, ranging from dubious to patently misleading, that recommend a variety of peculiar, and even extreme, diets. Such claims have a long history, but modern mass media have amplified these misleading recommendations.

The science of nutrition

As with so many other essentials of modern understanding, the science of nutrition took off during the latter half of the 19th century, made major advances (including the discovery of all vitamins) before the Second World War, and reached an even deeper and more complex understanding after 1950.[1] This stream of better understanding has been translated into a multitude of dietary recommendations, ranging from official documents reflecting the scientific consensus of the time to miraculous new diets. Detailed guidelines for recommended intakes of food energy, carbohydrates, fiber, fats (including specific fatty acids), and protein (including essential amino acids) have been set by extensive consultations guided by the World Health Organization (for worldwide use) and by national expert groups.[2]

Recommended quantities differ according to age, weight, and sex and they are also affected by levels of activity, pregnancy, catch-up growth, and by some chronic ailments. For example, a heavyset 30-year-old man (weighing 90 kg) engaged in strenuous physical activity would need up to 4,200 kcal/day, while a slim (50 kg) 70-year-old woman who spends most of her time at home doing light physical tasks can do with less than 1,800 kcal.[3]

Only a few countries have reliable and repeated studies of actual food intakes (measured by weighing all consumed food, rather than by relying on inevitably inferior recall of past consumption), but the FAO updates its detailed national food balance accounts every year, showing average per capita food supply—that is, the amount of foodstuffs and macronutrients available for distribution and purchase, not the average daily consumption, which is considerably smaller due to the intervening food losses.[4]

Quantifying consumption

The FAO's balance sheets show that the population-wide averages of daily food energy supply are far above any conceivable dietary needs for most of the world's (now increasingly less strenuously active) population: in affluent nations, they are well over 3,000 kcal/capita (in the US about 3,600 kcal and UK 3,300 kcal, with Japan at about 2,700 kcal being the only exception), but China (about 3,330) and Brazil (about 3,200) are also above 3,000 kcal/capita, with India, Bangladesh and Nigeria, Africa's largest nation, at 2,600 kcal/capita. Among populous nations only Ethiopia and Pakistan are just below 2,500 kcal/capita, indicating that a non-negligible share of the population is under- or malnourished.[5]

The FAO's latest estimate of the global population's share in these two categories is about 10 percent, with a range of 9.2–10.4

percent (depending on the assumptions used to calculate the total).[6] This is a major improvement compared to the share of just over 30 percent in 1970 and about 15 percent in the year 2000, but an unwelcome increase from 8.4 percent in 2019. About 55 percent of these roughly 770 million people are in Asia, and more than a third in Africa. Although an eventual eradication of this detrimental condition will require further increases of food production (or food imports) in several nations, the primary need is to improve access to food—that is, ensuring that the poorest segments of affected societies get at least the required nutritional minima.[7] Such policies are always profitable, as they reduce morbidity and mortality (particularly in childhood) and result in healthier adult populations.

An old rule is that an adequate supply of food energy provided in any specific mixed diet prevailing in a particular country or region should also provide adequate amounts of the three macronutrients—carbohydrates, proteins, and fats—and national food supply summaries confirm that. Macronutrient requirements are best conveyed as ranges of overall food energy intakes: carbohydrates should supply 45–65 percent of all food energy (with 50–55 percent as the optimum associated with the lowest mortality), fats 20–35 percent, and proteins 10–35 percent.[8] All of these requirements are easily met by the food supply available in affluent and middle-income countries, as well as by the world's two largest nations, China and India, and by the majority of populations in the low-income nations of Asia and Africa. African countries with the lowest per capita food supply include Madagascar, the Central African Republic, and Zimbabwe.

Macros

Given the typical makeup of low-income diets (recall this book's second chapter, on the dominance of grains), carbohydrate share is the easiest one to meet; the undernourished tenth of humanity is more often short of adequate protein and fat supply. While the weight-dependent levels of recommended daily protein intakes are approximately 40–60 grams per day for fast-growing teenagers, adults need only 0.8 grams per kilogram of their body weight, which means that most people do not require more than 45–60 grams a day. During pregnancy and lactation, women need about 70 grams a day.[9] Protein intakes should also be corrected based on their digestibility: cow milk and egg proteins are perfectly digestible; digestibilities are also high for meat proteins (0.95 for chicken), much lower (0.75) for beans, and only 0.4 for wheat.[10]

Clearly, relying only on a single source of low-digestibility protein is unadvisable, but normal mixed diets that supply enough energy also suffice to provide any conceivable protein needs. In fact, daily per capita protein supply is now not only much higher than necessary (100–110 grams a day) in all high-income nations (per capita gross national income above $13,206 in 2023), it is also over 100 grams in China and about 130 grams in Brazil (both in the middle-income category). India and Nigeria just about meet the need (averaging 60 grams a day), and Bangladesh and Ethiopia have an inadequate supply with only around 30 grams a day. Dietary fat supplies show a similar distribution: well in excess of requirements in affluent nations as well as in China, Brazil, and Mexico; too low in Nigeria, Ethiopia, and also in Bangladesh.

Nutritional advice: take it with a pinch of salt

After attaining more comfortable levels of income, people begin to eat more selectively. Reduced intake of pulses has been the universal sign of modern food transitions, with India, clinging to its dal, as the notable exception. Higher meat and sugar consumption has accompanied nearly all disposable income increases around the world, as has the eating of fresh fruit.[11] These gradual transformations of food intakes may result in diets that are suboptimal—or even outright undesirable—for maximizing the health benefits of eating and, ultimately, increasing life expectancy. Or so have many students of nutrition claimed, as they condemn diets heavy on many specific nutrients and foodstuffs and extol other, often equally one-sided, recommendations.

A complete enumeration of these claims would be tedious, but even cursory observers of dietary news in the last 50 years could not have missed some of the leading recommendations and conclusions—some presented as results of extensive nutritional and medical research, others offered through mass media by self-proclaimed experts taking advantage of human gullibility and widespread desire to be healthier, slimmer, and live longer. The post-1950s world has seen many new, much-advertised (and commercially exploited) diets whose competing claims stress their supposed scientific foundations and uniquely admirable outcomes.[12]

Extremes

That two much-touted choices—no animal foods ever (driven by health as well as by ethical concerns) and a highly meaty diet (promoted to improve health)—have been complete opposites tells us as much about the underlying expertise as it does about the propensity to becoming a true believer. Veganism does not allow even a drop of milk (that we are mammals and hence well suited to consume mammalian milks is seen as completely irrelevant) or a single egg (albeit laying it certainly does not involve killing the bird, and for a free-ranging bird there is also no unnatural confinement), and absolutely excludes any meat and fish. Vegetarian diets are far more forgiving, ranging from lacto-vegetarianism (dairy products allowed) to an all-but-terrestrial-meat diet (lacto-ovo-pisci vegetarianism).

The evolution of our species and its specific nutritional requirements (particularly during pregnancy, infancy, childhood, and puberty) offer no foundations for recommending veganism as a population-wide choice—but it remains a dietary option for adults with access to adequate sources of all plant-derived nutrients.

In contrast, all diets with reduced meat consumption in countries where substantial meat intakes have been the norm, are to be strongly recommended, especially for adults, on a nutritional basis, and they would be also more acceptable, and easier to adhere to, by people who had previously followed mixed and highly meaty Western diets.[13] At the same time, their environmental benefits might be lower than is commonly thought because the additional production of highly nitrogen- and water-intensive nuts, fruits, and vegetables, as well as the now-common long-distance transportation of these foodstuffs, may partially negate the benefits of producing less animal feed.

The extreme opposite of veganism is the misleadingly named Paleolithic diet, which advocates eating plenty of meat (of all kinds) supplemented by vegetables and fruits. (Very few of our ancestors constantly ate diets consisting almost solely of meat.) Both extremes represent impractical templates for mass-scale adoption: their institutionalization and enforcement would require a radical transformation of modern food production—on the one hand abolishing large productive sectors (meat, eggs and dairy, fishing and aquaculture) that sustain many economies (with profound economic consequences), on the other hand a mass-scale expansion of meat production with the intensification of all (already outlined) accompanying downsides. And, of course, in both cases an unprecedented population-wide reversal of long-established dietary habits!

If we were all Paleo

How these shifts could be achieved to make a notable difference on a national or global scale remains unexplained. There can be no doubt that the promoters of highly meaty diets are especially delusionary and that they have not thought about how to supply the requisite amounts of meat without further massive deployment of the energies and material resources required to multiply the annual production of meat. In 2020 global meat consumption supplied only about 8 percent of all food energy and 17 percent of all dietary protein.[14] What would be the environmental consequences of at least tripling the worldwide animal slaughter in order to supply a quarter of all energy and half of all protein in highly meaty (but still insufficiently "Paleolithic") diets? The next step would be to calculate by how many times our output of animal feed (forages, grains, tubers, legumes) would have to increase: the result would depend on the composition of this enormous meat supply.[15]

If we were all vegan

The global adoption of a vegan diet carries its own problems. If the supply of a purely vegan diet was simply a matter of adequate food energy, then everything would be easy: we could just grow sugarcane, the plant with the highest carbohydrate yield per unit of land in a suitably warm climate.[16] But as explained in the second chapter, healthy nutrition is a matter of proper nutrient balance, and while a planet of vegans could easily be supplied with staple cereal carbohydrates (also containing protein) and plant fats, providing plenty of high-quality protein would be the greatest challenge.

Inevitably, this would mean the ascent of grain legumes. In traditional Old World societies, pulses (peas, beans, and lentils in Europe; beans and peanuts in Africa; lentils, beans, and soybeans in Asia) were second staples, as their annual per capita consumption was as high as 25 kilograms in India and more than 10 kilograms in many countries in Latin America and among poorer populations in Europe.[17] Those intakes provided 15–30 percent of all dietary protein and up to 15 percent of all food energy. Increased affordability of meat, eggs, and dairy eventually reduced legume consumption to inconsequential levels not only in most of Europe (Germany's supply is now less than 1 kilogram per capita a year, France's less than 2 kg) but also in Japan (only about 1.5 kg) and China (less than 1.5 kg).[18]

Those rates mean that less than 1 gram of protein, or no more than about 1 percent of daily protein intake, is contributed by legumes—and that a tenfold increase of consumption would be needed even if pulses were to supply just 10 percent of all protein in vegan diets. Such a consumption shift is not going to happen: well-known downsides of copious pulse consumption (soaking, protracted cooking, problems with digestion, and lack of gluten, which makes them unsuitable for bread and noodles) do not make beans, peas, lentils, and soybeans the mainstream, time-saving choice in modern societies preferring quick food preparation and easy digestibility.[19] This means that among populous countries, relatively high legume intakes will remain limited only to India (about 15 kilograms per capita a year) and Brazil (about 13 kg)—where, respectively, lentils and beans (dal) and black beans (*feijão preto*) remain long-cherished national staples.[20]

Substitutes

There are two alternatives to eating legumes. The first one is processing soybeans to make such traditional products as bean curd (Japanese *tōfu* and Chinese *doufu*, prepared by the wet-milling of soybeans and then coagulating them with calcium or magnesium sulfate), miso (soybeans fermented with rice or barley and *Aspergillus oryzae*), and soy sauce (milled soybeans and wheat fermented with *Aspergillus sojae*).[21] The second one is using legumes to prepare textured plant proteins (fibrous structures resembling the appearance and mouthfeel of meat) as the basis of so-called meatless meat.[22] During the last few years it has been impossible to miss the numerous superlative claims about the disruptive and transformative nature of this shift, made by the expanding meatless meat (and alternative eggs and also fish) industry that claims to improve diets even as it reduces the risk of global warming.[23]

In some countries with traditionally high animal food intakes, meat is hard to dislodge: reduced consumption of beef has been common in the Western countries (made up largely by eating more chicken), and reduced intake of all kinds of meat can be seen in many national statistics, but these trends do not mean that meat has been replaced by alternatives. Bean curd has been available in North American supermarkets for decades, but Americans and Canadians have not abandoned processed meat products in favor of that soybean derivative. Tofu has not displaced pepperoni on pizza; it has hardly made even the proverbial dent. Every year North American per capita consumption of tofu is lower than Japan's average daily rate![24] And despite the recent waves of plant-based "meat" offerings (and a spike of plant-based meat-mimicking sales during the first pandemic year), meat substitutes have a very long way to go before making a real difference. US sales tell the story: in 2021 the country spent nearly $160 billion on meat (nearly $120 billion on fresh-cut, and the rest on processed meat products) while tofu sales were about $350 million and the value of all sold meat substitutes was $1.3 billion (less than 1 percent of all meat sales); by 2022 the sales of substitutes rose to $1.8 billion but their total volume had declined, and even the best growth expectations would not raise their value above 10 percent of meat sales by 2030.[25]

Planet of the plant-eaters

But if populations raised on great food variety had to go strictly vegan, they would certainly demand more than a high-bean diet: meatless meals would contain not only more legumes and cereals to provide the three macronutrients but also more vegetables, fruits, and (for concentrated protein and fats) nuts. Because of the elimination of animal feeds, expanded cultivation of food legumes and staple grains would not cause any additional land or input claims, but vegetables, fruits, and nuts have much higher labor input needs than staple field crops and their expanded cultivation would increase many already obvious demands.[26]

Vegetables grown outdoors have high fertilizer and water demands, because many crops are now grown and harvested in the same year consecutively: in California, the production and harvesting period extends for 12 months for vegetable crops ranging from beets and broccoli to radishes and spinach, as well as for strawberries.[27] While a single crop of vegetables will average 2–3 times more water per unit of energy than a staple cereal, there may be three consecutive crops grown every year in the same field (broccoli, kale, and lettuce mature in 60 days; artichokes, cabbage, and cauliflower in about 100 days).[28]

And vegetables grown in greenhouses—the Dutch (in glass) or Spanish (under plastic tunnels, especially in Almeria) model, but now increasingly across Asia and North America (and not only for tomatoes, peppers, and cucumbers)—have extraordinarily high requirements for energy in order to build the structures, to heat and light them, to pump water, and to harvest, process, and distribute the products.[29] As a result, a greenhouse tomato may need energy inputs an order of magnitude higher to produce the same amount of vitamin C (the nutrient that makes a tomato nutritionally valuable) as does the field cabbage![30] And nuts, so highly recommended in vegan literature, have (as already noted in the third chapter) no equals for their high water demand, as they consume 6–7 times more water per unit of food energy than cereal staples.[31]

Restrictive diets, empty claims

There have been other, less sweeping dietary recommendations that have stressed specific components of diets. One specific recommendation that has been supported by subsequent studies is to eat plenty of indigestible fiber in order to prevent colorectal cancer.[32] Diets dominated by one specific macronutrient have included not only the just-noted high-protein, low-carbohydrate diet heavy on meat but its very opposite: a high-carbohydrate diet promoted for weight loss and lower risks of heart disease.[33] Other diets aim at maximizing intakes of specific compounds, with antioxidants being particular favorites. In the extreme, this would lead either to a purely fruitarian diet or to swallowing daily capsules of fish oil.[34]

Of course, none of these diets or supplement regimes should be judged by the often uncritically effusive and sometimes patently ridiculous claims that their proponents advance on their behalf. Their effectiveness should be compared with the already noted, well-established recommendations for daily nutrient intakes and with the findings of large-scale, well-designed, long-term nutritional studies. And there should be no rush to judgment: wait another decade, and the scientific consensus may have shifted rather significantly, and the original claims may look questionable.

The bloated story of dietary fats

I will illustrate these reversals of dietary fortunes by focusing on the role of dietary fats in cardiovascular disease (CVD) mortality. CVD has been the leading cause of premature death in all affluent countries, and hence any effective intervention would be much welcome.[35] The first dietary recommendations resulted from the pioneering Framingham Heart Study, initiated in 1948–50 and still continuing: reduce the intake of saturated animal fats (solid at room temperature) and cholesterol (consumed separately in butter or lard, or as a part of fatty meats and dairy products).[36] They were amplified and popularized by Ancel Keys, an American scientist who studied chemistry, economics, and zoology before he turned to nutritional research. In 1958 he launched the Seven Countries Study of diet and CVD, which resulted in the total condemnation of diets high in saturated fats and also highly recommended the traditional Mediterranean diet characterized by high intakes of fruit, vegetables, and fish, and the restrained consumption of meat.[37]

More specifically, Keys and his followers advocated the radical replacement of saturated fats by plant oils—mono- (olive) or polyunsaturated (sunflower, rapeseed, peanut)—but not by plant-based hydrogenated oils (margarine).[38] New dietary guidelines reflected these recommendations, as high-income countries in general, and the US in particular, pursued a public campaign against saturated fats. But the French exception (a diet high in saturated fats coexisting with relatively low CVD mortality) has always suggested that matters are not that simple.[39] And so they proved to be. New studies, with large numbers of participants and long durations, weakened or even subverted the "bad fat" narrative, finding that consumption of saturated fats did not lead either to higher mortality in general or to increased CVD death rates in particular—and, moreover, that the consumption of monounsaturated and polyunsaturated fatty acids did not have an overtly beneficial effect.[40] And eventually, new US dietary guidelines did away with the long-standing advice to limit cholesterol intake to less than 300 mg/day.[41]

We now appreciate that the effect of specific foods on coronary heart disease does not depend just on their content of saturated fats, because individual saturated fatty acids differ in their impact.[42] Moreover, the outcomes of dietary interventions aimed to reduce CVD mortality depend on the kind of macronutrients that take the place of saturated fats. Most notably, replacing saturated fats with a higher carbohydrate intake (particularly refined foods) can be counterproductive.

Consequently, the focus should be on the overall composition of diets rather than on individual foods or their components.[43] Or as Frank Hu, an American physician and nutritionist with a long record of dietary studies, has noted, "the single macronutrient approach is outdated . . . future dietary guidelines will put more and more emphasis on real food rather than giving an absolute upper limit or cutoff point for certain macronutrients."[44]

Real food

For decades, and long before so-called ultra-processed foods became the latest villain in nutrition (a passing fad: just another name for too much fat, sugar, and salt), I have been advocating a real food approach: we need to look at the ultimate outcomes of lifelong food intakes, i.e. at the frequency of diseases that are known (or suspected) to be connected to improper nutrition, and trace the changes of average life expectancy. This is bound to be more revealing than concentrating on time- and population-limited studies of specific nutrient intakes or mortalities. Its most convincing version is to look at the ultimate outcome: life expectancy at birth. This variable is influenced by a multitude of factors (genetics, preventive healthcare, general lifestyle, nutrition, chronic ailments), but the expectancy cannot be high if the lifelong food intake of a population has any chronic detrimental effects.

After the Second World War, all affluent countries benefitted from very similar levels of healthcare (preventive and acute) and a more than adequate average food supply, and they also experienced a convergence of lifestyles (higher car ownership, more sedentary work). As a result, until some reversals due to the Covid pandemic, life expectancies in the EU, US, Canada, Australia, and Japan have been uniformly increasing—and, moreover, rates of this improvement have been similar in countries whose diets are very dissimilar.[45]

Japan vs. Spain

The best example of this reality is the comparison of Japan and Spain. In 1950, average (combined male and female) life expectancy was almost identical: 60.64 years in Japan and 60.21 years in Spain. Seventy years later both countries, after experiencing significant dietary and lifestyle transformations, gained more than 20 years of average life, but Japan is about a year ahead: 84.67 vs. 83.61 years. And within the EU the longest combined life expectancies in 2020 were in Spain, Sweden, and Italy, all with 82.4 years. France was insignificantly behind with 82.3 years, but it led in female longevity (85.3) ahead of Spain (85.1) and Italy (84.7), while Sweden had the male primacy (80.7) compared to Italy's 80.1 and Spain's 79.7 years.[46]

And yet Spain's great life extension has been accompanied by one of the most far-reaching, and relatively rapid, dietary transformations—much of it including a shift that is often seen as undesirable. The shift began during the 1960s and accelerated in 1975 after Franco's death, then again in 1986 on accession to the EU. Between 1960 and 2000, Spanish per capita meat supply nearly quadrupled, the supply of animal (saturated) fats tripled, and consumption of dairy products rose by a quarter—while the intakes of olive oil and cereals declined, as did the drinking of supposedly cardioprotective wine.[47] The rise in meat consumption was so rapid and so substantial that, by the end of the 20th century, Spain was the EU's most carnivorous country, with per capita supply (in terms of carcass weight) averaging more than 110 kilograms a year—compared to, respectively, 100, 80 and 70 kilograms a year in such traditionally meat-eating nations as France, Germany, and Denmark.[48]

The best explanation of this paradoxical outcome is a combination of factors, including increased consumption of fish and fruit and, notably, better access to preventive medical care and a reduction in smoking.[49] The latest studies show that the rate of the Spanish decline in CVD mortality slowed from −3.7 percent (men) and −4.0 percent (women) between 1999 and 2013 to −1.7 percent and −2.2 percent after 2013. This is in accord with the slowdown in other affluent countries, and the most likely explanation is that the current high longevity averages are approaching the expected population-wide maxima of life expectancy.[50]

China offers an even starker—and much larger-scale—example of a major dietary transition marked by large increases in per capita rates of meat and saturated fat supply, accompanied by steady gains in average life expectancy. Between 1970 and 2020, meat supply rose nearly eightfold and in many cities is now nearly at the European level.[51] This increase was dominated by traditionally preferred (and often fatty) pork consumption, and hence the supply of animal fats roughly quadrupled in half a century, while average life expectancy increased from 60 to 77 years, closing in on the American pre-Covid mean of 78.8 years in 2019.[52]

What should we change about our diets?

Concerned consumers can easily check, and follow, the latest national dietary recommendations, but these comparisons of diets and longevities show that, different as they are in many particulars, the prevailing compositions of typical diets in countries as different as Spain, France, and Sweden or Japan and China are all compatible with efforts to prevent excessive CVD mortality and to prolong life. There is no need for any radical interventions, least of all for following any extreme diets or for advocating regular megadoses of specific dietary supplements—be they vitamins, minerals, or fish oils.[53] But this does not mean that there is no need whatsoever for some gradual dietary modifications, as well as for some immediate, more deliberate interventions.

The first need, and the one that could be accomplished with relatively little expense, is to ensure that large shares of populations—above all children—in many low-income countries are not suffering due to several avoidable micronutrient deficiencies. In comparison to other major nutritional challenges, micronutrient deficiencies can be alleviated by using readily available, highly effective, and inexpensive measures. The fortification of foodstuffs and the distribution of supplements are needed in all cases where adequate levels of essential vitamins and minerals cannot be provided by the prevailing diet.[54]

Many people are not even aware of this practice, but some fortification measures began many decades ago—the iodization of table salt in Europe during the 1920s, the fortification of wheat flour in the US and Canada with iron and four of the B vitamins (thiamin, niacin, riboflavin, and folic acid) in 1941—and they have since expanded worldwide. Without any doubt, food fortification ranks as one of the most cost-effective public health interventions: only vaccination—particularly the 5-in-1 vaccine that protects children from diphtheria, pertussis, tetanus, and two forms of hepatitis—yields a greater return on its relatively modest investment.[55]

Unfortunately, even in the early 21st century, micronutrient deficiency disorders remain a major public health problem worldwide (even in high-income countries). They affect all groups of people, but their worst impact is evident among the children and women in the poorest African, Asian, and Latin American nations. The most common deficiencies are in vitamin A (fish, eggs, dairy), iodine (a trace element in seafood and dairy), and two common metals, iron and zinc (high in meat and dairy products).[56] Assessing the worldwide prevalence of these deficiencies requires representative sampling and blood testing; not surprisingly, in many low-income countries such data are both scarce and outdated, and hence the worldwide totals of micronutrient deficiencies are not known with a high degree of accuracy.

About 250 million people do not have enough vitamin A, and that includes more than half of all children in more than half of all sub-Saharan countries in Africa.[57] This deficiency affects eyesight (night blindness, and progressive xerophthalmia leading to blindness) and the immune system, raising the chances of illness and premature death. Regular consumption of carrots, squashes, sweet potato, liver, fish, and cheeses eliminates the risk, improves immune function, and reduces mortality—above all, due to measles and diarrhea—as does the fortification of oils and fats and regular supplementation. Successful vitamin A supplementation programs for infants and children up to five years of age have been undertaken by many countries, but a recent study found that most data on the actual prevalence of deficiency are outdated or absent (in Africa's Sahel countries, and also in India and Kazakhstan), which means that the actual shortages are even more common. That study also found that the deficiency is severe in more than 30 countries, with prevalence in excess of 30 percent.[58]

Iron deficiency

Iron deficiency is the leading cause of anemia, a low count of the red blood cells that distribute oxygen to all body tissues. In 2019,

1.76 billion people—which is about 23 percent of the global population—were anemic, with about 55 percent of the total having mild, about 40 percent moderate, and the rest in the severe category.[59] Regional rates ranged from less than 5 percent in Western Europe to 20 percent in tropical Latin America and to more than 30 percent in South Asia and most of the countries of sub-Saharan Africa, with Zambia (49 percent), Mali (47 percent), and Burkina Faso (46 percent) having the highest prevalence.[60] Iron deficiency remains the leading cause of anemia, and children and young females in low-income countries remain the most affected populations. Clinical signs and symptoms range from poor mental performance (if untreated in infancy, it may result in lifelong cognitive challenges), intolerance of cold, and shortness of breath. Restless leg syndrome and pica (the compulsive eating of non-food items) are also common manifestations.[61]

Women (because of menstrual blood loss) and children (because of their higher iron needs) are more vulnerable. Moreover, a recent finding indicates that the current test thresholds for detecting iron deficiency in women and children may be too low (detecting only the more severe forms), and that undesirable changes take place at levels higher than the currently defined thresholds. Given these new thresholds, even in the US about 30 percent of women and children could be iron-deficient, while according to existing standards the prevalence is about 17 percent among premenopausal women and 10 percent among children.[62] If extended to the entire global population, this would at least double the total of people suffering from iron deficiency, making a case for more of the obligatory fortification of staples (now compulsory in more than 80 countries) and for more supplements.

Iodine deficiency

On the global scale, iodine deficiency is about as prevalent as the inadequate intake of iron: the best estimate is that about 2 billion people are affected, including more than 200 million children.[63] A similar number of people have clinical manifestations, including some 50 million people with goitre—the swelling of the thyroid gland. Screening programs for early iodine deficiency are especially important among pregnant women, because iodine deficiency during pregnancy increases the likelihood of miscarriage, stillbirth, infant mortality, and birth defects. During infancy it impairs both physical and mental development, depriving the affected children of ever having normal, productive lives.

The solution is a universal use of salt fortified with iodine, using at least 15 parts per million. This is an inexpensive measure, but achieving global distribution has not been easy. Universal salt iodization programs have been proposed in most countries, but so far they have covered no more than 75 percent of the world's population. Even India, a country that began introducing iodization during the 1980s, had not achieved complete coverage three decades later.[64] Not surprisingly, iodized salt is not readily available in many parts of Africa affected by deep poverty, civil conflicts, and recurrent natural disasters.

Zinc deficiency

Because zinc is present in hundreds of enzymes and many other proteins, it is indispensable for metabolism, for the growth and differentiation of cells (hence for normal pregnancy), for cell-mediated immunity, and for the development of the nervous system.[65] The most common manifestations of zinc deficiency are poor child growth, increased childhood morbidity and mortality, reproductive problems, and reduced immunity. The prevalence of zinc deficiency has not been studied as much as the extent of other micronutrient shortages. Older estimates suggest that some 30 percent of the world's population were affected early in the 21st century; more recent estimates put the global total at 17 percent with the highest rates in Africa, where a quarter of the population may bear the consequences and where it contributes to stunted growth.[66] Some national studies indicate far higher prevalence.

Serum samples from the Ethiopian National Micronutrient Survey conducted in 2015 indicated that 72 percent of the country's population was zinc-deficient, with high prevalence in all population groups.[67] Fortification of some foods and inexpensive supplements prevent deficiencies.

While it would be naive to think that all of these deficiencies could be eliminated before 2050, it is quite realistic to expect further worldwide reduction of their prevalence, with some affected countries reducing the shortages to fractions of their current extent. Food fortification programs, now active in scores of countries, can be very effective—but, as with any large-scale continuous effort, their success depends on putting in place appropriate production and distribution capacities, oversight, and monitoring. Obviously, these conditions are lacking in places beset with endless conflicts, poor governance, and weak service infrastructures. And, not surprisingly, multiple micronutrient supplements are more cost-effective than addressing deficiencies separately.[68]

Hungry and undernourished

An objective appraisal indicates that dealing with the second-most pressing challenge—to provide enough food for the roughly 10 percent of the world's population that suffers from hunger and undernutrition—should not be more difficult than addressing micronutrient deficiencies. This conclusion rests on two facts: the total number of people affected and the dominant cause of these shortages. First, the total number of people experiencing chronic food shortages (and this takes place not only in low-income countries) is comparable to the total afflicted by micronutrient shortages. The FAO estimates that between 720 and 811 million people "faced hunger" and nearly 2.37 billion people "did not have access to adequate food" in 2020. Second, this is not primarily due to actual food shortages but to "persistently high levels of poverty and income inequality."[69]

That the reduction (or even complete elimination) of inadequate food intakes is not primarily a matter of increased production is best illustrated by the existence of non-negligible undernutrition among low-income groups in many affluent countries. Average per capita food supply in all high-income nations is far above any conceivable need—and yet undernourishment and food insecurity are found even within these countries, as *access* to food matters more than its actual availability. Even in Canada, one of the world's leading food producers and major food exporters, some 17 percent or 1.2 million children live in food-insecure households.[70]

The United States has dealt with these shortages through such measures as school breakfasts and lunches, and by issuing food stamps, and increasing food insecurity led France to consider giving payments to poor families to compensate for rising food prices.[71] Such formal measures are absent in the countries that would need them most, and their absence has led to proposals for establishing a system of global food stamps that would be linked to particular problems experienced by the most vulnerable populations (the elderly; women with young children), perhaps with eligibility and distribution combined with other measures of economic disadvantage.

Hope in sub-Saharan Africa

The greatest hope, and the greatest frustration, lies in the fact that sub-Saharan Africa, the region most in need of such measures, also has the highest potential to improve typical crop yields. Higher yields would boost local supplies, reduce (in many countries, rising) dependence on imports, and improve accessibility to staples. This region has the world's highest yield gaps: the difference between the global harvest mean of a specific crop and its national level. Recently, average staple yields in Nigeria, the continent's most populous nation, have been far lower than even in chronically food-short Ethiopia (rice, about 1.4 vs. 2.4 t/ha; corn, 2 vs. 3.3 t/ha) and a fraction of averages in Brazil (4.6 t/ha for rice, 5.1 t/ha for corn).[72] As a result, while India has recently been importing less than 0.5 percent of the staple cereals it consumes (and China's share was about 3 percent), Nigeria buys 15 percent of its staples abroad.[73] Moreover, Nigeria, formerly a major food exporter, now also needs to import oilseeds, while until the 1960s Kano, its northern state, was famous for its impressive groundnut pyramids, each made of up to 15,000 bags of harvested peanuts ready for shipment.[74]

A large part of this sub-Saharan underperformance is attributable to very low intensities of cultivation, and above all to limited mechanization and irrigation and to very low rates of fertilization: recent applications averaged only about 3 kilograms of nitrogen per hectare of agricultural land, compared to more than 30 kilograms in Europe and about 50 kilograms in China.[75]

At the same time (as I will stress in the final chapter), sub-Saharan Africa does not have any regions whose soil quality would resemble the US Corn Belt, the Argentinian pampas, or the Ukrainian-Russian belt of black soils (chernozem)—unlike those relatively young, deep soils rich in organic matter, African soils are old, leached, and inherently much less fertile.

This natural constraint has been made much worse by the fact that sub-Saharan Africa has the least stable conditions to pursue the goal of expanded food production in a steady and effective manner. Among Asia's 48 countries, only nine have a political stability index below −1.0, and 16 have it above 0 (higher values indicate more stable countries); while of 47 sub-Saharan African countries, only nine have an index above 0, and 13 rank below −1.0, including populous Mozambique, Congo, Ethiopia, and Sudan, whose combined population of 485 million people in 2020 was larger than that of the EU. Nigeria ranks seventh from the bottom, surpassed only by Congo, Sudan, Ethiopia, the Central African Republic, and Somalia.[76]

The region's food problem is thus a complex combination of natural constraints and social instability (above all, recurrent civil and cross-border conflicts). In contrast, Asia has succeeded in shifting all but a relatively small share of its huge population to the category of adequate food supply. By far the greatest achievement of

China's post-1980 economic modernization is not its ability to flood the world with cheap manufactures, but its capacity to feed 1.4 billion people while importing less than 10 percent of its food supply (Japan and South Korea import more than 60 percent of their domestic consumption!).[77]

What does the future hold?

Unfortunately, the immediate outlook is not for any rapid changes: the FAO anticipates that average per capita food energy supply in sub-Saharan Africa will improve only by about 2.5 percent by 2030.[78] This is particularly worrisome because the prevention of childhood malnutrition is one of the most cost-effective ways to ensure future economic growth.[79]

But important dietary changes do take place across generations, and they do result in new levels of typical consumption: the post-1950 shifts in affluent countries (some desirable, others not so much) included declining beef and rising chicken consumption (in per capita terms in the US, red meat was down 17 percent between 1960 and 2002, chicken up nearly 3.5 times!), drinking less milk (both in the US and in most of Europe, but eating more yogurt and more pizza cheese), the already noted decline of leguminous grains from secondary staples to marginal intakes, and shifts from butter to plant oils. As always, long-term gradual changes cannot be reliably predicted, but it would be surprising if the next three decades did not see adjustments as substantial as those of the past 30 years. And, quite possibly, they can go even further, and they may make an important contribution to feeding the world by the mid-21st century. There are too many exaggerated expectations for new, radical "solutions" in the modern world—it is gradual gains that will matter more.

7. Feeding a Growing Population with Reduced Environmental Impacts: Dubious Solutions

As we look a generation ahead and contemplate the state of the global food system in the middle of the 21st century, it is clear that besides continuing with many gradual improvements and adjustments of crop and animal production, it would be helpful if there were a few radical transformations able to satisfy two overriding, linked goals: to deliver adequate nutrition to the still-increasing global population, while reducing the inputs into the global food system and lessening its numerous environmental impacts. Doing more with less—that is, doing most of the things better—might be an alternative encapsulation of this book's last two chapters. At this point, given our understanding of food requirements and the levels and composition of actual food supply, the real challenges aimed at improving the global food system fall into three distinct categories, which we will look at in ascending order of difficulty.

1. Expanding existing food production to accommodate the nearly 2 billion people that will be added to today's population by the middle of the 21st century.
2. Reducing the indefensibly high level of food waste all along the food chain and in both high- and low-income countries.
3. Restructuring the global food system to reduce its multitude of environmental burdens.

Forecasts and accomplishments

The first task has a large regional overlap with the previously noted challenge of undernutrition. The UN's medium population projections see the addition of about 1.9 billion people between 2020 and 2050, with 60 percent of that total taking place in Africa and just over 50 percent in its sub-Saharan part.[1] That region will thus face a twofold task: producing more food to reduce malnutrition in the existing populations, and feeding another billion people. The situation will be very different in Europe, where the projection sees absolute population decline (the continent will grow only if it permits substantial immigration), North America (about 15 percent higher), and Asia (about 14 percent higher). Following Japan and South Korea, China's population has also stopped growing; India has become the world's most populous nation but a recently accelerating slowdown of its total fertility rate may cut future growth quite a bit.[2]

I will not try to describe the likely state of the global food system of 2050. As with any long-range forecasting efforts, it would be, at best, a mixture of a few fairly accurate conclusions and many major misses. Just think what even the best-informed assessments done in 1990 would have missed when describing the global food situation in 2020. By far the most important developments that would not have been anticipated in 1990 were the changes in China, India, and Russia. The economic rise of China has entailed an unprecedented three-decade run of growth (with a multitude of domestic and foreign consequences), and it has led to the country's average per capita food supply rising by a third, to a level that is only about 5 percent short of the averages in France, Germany, and Italy, and far above the Japanese mean.[3]

Indian accomplishment has been almost as impressive. Although the country's average per capita food supply mean is still about 20 percent behind China's (now clearly excessive) rate, it has risen by nearly 20 percent since 1990 even as the country added 507 million people—that is more than the total population of the EU—and, until very recently, its imports of staple grains remained negligible.[4] And in 1990 the USSR remained a notorious agricultural underperformer and grain importer; it bought about 33 million tons of cereals. But by the end of 1991 the state had ceased to exist, and in 2020 Russia, after becoming the world's leading wheat exporter, sold 43 million tons of cereals abroad.[5] Even more recently, the short-term impacts of the Russo-Ukrainian War have been well covered, and it is too early to make any long-term verdict on the global food supply.

The short term

The most likely trajectory of the global food system until 2030 is much easier to describe. The FAO in its projection rightly expects that the growth in agricultural production needed to feed a growing global population will come overwhelmingly (nearly 90 percent) from further productivity improvements—that is, without further expansion of cultivated land.[6] Large yield gaps indicate that expecting further substantial gains in many low-income countries is quite realistic, but this will require continued investments in inputs and infrastructure. If African crop yield gaps are not reduced, it will not be because of environmental obstacles or agronomic insufficiencies but because of continued conflicts and poor governance. The efficiency of animal feeding should improve in most middle-income countries that are expected to experience greater demand for meat, eggs, and milk, and well before 2030 aquaculture should top capture fisheries as the main supplier of protein from the sea.[7]

High-income countries have no need for any further food production increase; they need just the opposite. A deliberate reduction of the most intensive methods of cropping and animal husbandry (exemplified by the already unfolding limits on Dutch intensive farming) and moderation of meat consumption should ease some of the worst environmental impacts in the EU, North America,

Australia, and Japan. Two major developments may lead to notable shifts in agricultural output in high-income countries: the future extent of biofuel production, and possible limits imposed by measures designed to slow the pace of global warming. In the US, 40 percent of grain corn's annual harvest has been converted every year to ethanol since 2013; in Brazil, 55 percent of all sugarcane was used in the same way in 2021. These shares may grow, and the practice may be extended to other countries with the growing use of biofuels by airlines—or it may be reduced as an aggressive adoption of electric cars lowers the demand for automotive fuels and additives.[8]

Beyond forecasts: questionable efforts

Perhaps the best way to look beyond 2030 (without offering dubious forecasts and necessarily flawed scenarios) is first to review a short list of recently extolled measures that are unlikely to work or that might succeed only in limited ways, performing much below the exaggerated expectations that have been recently repeated not only by mass media but also by many enthusiastic business and investment promoters. The book's final chapter will look at those steps that *could* bring productivity improvements even as they reduce waste and the environmental impact of the global food system.

During the first two and a half decades of the 21st century we have been subjected to a flood of claims about scientific breakthroughs and engineering innovations, often seen as no less than disruptive, transformative, and even epochal. Given this uncritical enthusiasm, now prevalent in the news and in popular books (by non-experts and "experts" alike), it is important to offer a measured counterweight, a realistic assessment of those proposals, measures, and transformations whose novelty and promise have attracted a great deal of attention but whose practical contributions to changing the global food system by 2050 will, most likely, remain limited.[9]

I will address four of these major transformations (radical departures from prevailing practices) which have been proposed to reduce—if not eliminate—the negative consequences of prevailing practices even as they assure an adequate supply of food:

1. The universal conversion of food production to organic farming—a reiteration of a persistent notion about the form of an ideal (natural) food-producing system that would do away with the ills of modern agriculture.
2. Large-scale adoption of permacultures and polycultures to create a new way of cropping, bringing obvious labor and input savings while also reducing soil degradation and improving water retention.
3. Genetically modified staple crops that would be able to supply their own nitrogen needs or convert solar radiation with significantly higher photosynthetic efficiency, greatly reducing food production's environmental footprint.
4. Mass-scale production of cultured animal foods, above all *in vitro* (or, more accurately, in stainless steel) meat, but also fish, eggs, and milk, which would end up eliminating domesticated animals save for a small number of cell donors, and also do away with the need for producing a large share of crops grown for animal feed.

Organic farming: for the few, not the many

Perhaps the most radical of all unlikely solutions aimed at securing enough food by the mid-century is to reverse the trend of the past 150 years and convert the entire planet to organic farming. As it is usually defined, this practice means doing without any synthetic fertilizers, and that means relying on the recycling of crop residues and manures and on the planting of leguminous species to supply the necessary nitrogen—as well as not using any synthetic pesticides, herbicides, or fungicides, and relying instead solely on natural pest and weed control strategies.[10] How could one object if these radical shifts (no synthetic fertilizers, no other agrochemicals) took place without any negative effects on yields and the reliability of harvests?

But anybody familiar with the global nitrogen cycle and with crop requirements for the most important macronutrient—readers, this includes you—must have deep doubts about maintaining today's global harvest without any synthetic nitrogenous fertilizers. Even before their invention, producers sought to augment organic recycling by inorganic compounds in order to provide all plant macronutrients as needed. Chilean nitrates derived from guano deposits were first mined in 1826, though they did not become widely used until half a century later. After 1913, the Haber-Bosch synthesis of ammonia from its elements took over the world's nitrogen provision; large-scale potash mining began in 1861 in Saxony; and the extraction of Florida phosphates started in 1883.[11] When expressed in terms of pure nutrients, by 2020 global agriculture has been applying annually nearly 100 million tons of nitrogen, about 20 million tons of phosphorus, and 30 million tons of potassium.[12]

Synthetic nitrogenous fertilizers are needed in the largest quantities and make the greatest yield difference. Without their application we could not feed (assuming today's prevailing diets) at least 40 percent of humanity, and even if we were to recycle every bit of nitrogen-containing organic waste (crop and food processing residues; animal and human wastes) and expand the cultivation of leguminous plants (whose symbiosis with nitrogen-fixing bacteria puts additional nitrogen into farm soils) we could not match today's Haber-Bosch synthesis used to make solid (dominated by urea and nitrates) and liquid fertilizers.

And yet, some early published comparisons showed that organic cropping is nearly as productive or even surpasses the conventional ways, and one 2007 study flatly concluded that organic farming could supply enough food energy "for the whole human population eating as it does today" because "nitrogen-fixing legumes used as green manures can . . . replace the entire amount of synthetic nitrogen fertilizer currently in use."[13] Both of these conclusions are suspect, as they rely on questionable data regarding crop productivity and on simplistic assumptions concerning the mass-scale extension of leguminous cover crops (alfalfa, clover, etc.). More rigorous studies have shown that organic crops yield on average 20 percent less than the conventional ones, and that the challenges in maintaining nutrient supply in organic systems at rotation, farm, and regional levels would likely result in an even higher disparity.[14]

This was confirmed by the most recent meta-analysis of yield data and the intensity of soil use (years with a harvest crop in relation to the duration of rotation): organic yields were, on average, 25 percent lower, and the gap was 30 percent for cereals. Moreover, when combining the yield gap with the reduction in the number of crops harvested in the rotation, the productivity gap rose to 29–44 percent depending on the crops included in the rotation.[15] Some assessments have been even more emphatic, concluding that "organic agriculture cannot feed the world, because there is substantial scientific evidence that crop yields are considerably lower in organic systems. The long-term yield reduction could be as much as 40–50 percent compared with the corresponding conventional crops. Therefore, to obtain equivalent yields in organic systems, significantly more land would be needed for agricultural crops. However, according to recent assessments, such land is not available in the world."[16]

Germany provides a perfect illustration of the reality. The country has been, not surprisingly, a leading proponent of this radical shift, aiming for 30 percent of arable land to be farmed organically by 2030, but in January 2023 a new, detailed German study (prepared by a team of scientists at the Technische Universität München) concluded that, while organic farming has many (especially environmental) advantages, it requires significantly more land than conventional cropping: comparisons of "ecological" and conventional operations show nearly 2.3 times more land for winter wheat and twice as much for silage corn.[17] Yet at the same time, Germans are being urged to eat less meat in order to free up more land for planting the biofuel crops required to achieve another green target—that of eliminating fossil fuels in combustion.[18] Where is all that extra land going to come from?

Organic miscalculations

Nitrogen is the critical input, but in order to conclude that leguminous cover crops could provide more nitrogen than was applied in synthetic fertilizers at the beginning of the 21st century, the authors of the 2007 global organic farming study assumed that the planting of green manure crops would expand from about 11 percent of total cropland to all of it—that is, to some 1.5 billion hectares of arable land. But that is purely an academic exercise; when making this assumption about growing two crops a year on all of the world's fields, the authors were obviously unaware of at least five realities that make such a mass-scale transformation impossible—or, at best, most unlikely for decades to come.

First, a significant share of the world's cropland is already cropped more than once a year. Double-cropping is limited in the US but is common in parts of Asia, Latin America (much of Brazil's Cerrado is double-cropped with soybean and either corn, sorghum, or cotton; much of northern Argentina produces three crops in two years), and Europe. In the early 21st century it was practiced on about 12 percent of global cropland, with 34 percent of rice and 13 percent of wheat grown that way, and with China double-cropping 34 percent of its cultivated land (mostly rice after rice) and triple-cropping 5 percent.[19] Moreover, in intensive vegetable production (no matter if in California or in Java) as many as four or five crops are grown in the same soil every year.

Second, a similarly large share of the world's cropland lies fallow: it is not cultivated either during a single crop-growing season or for an entire year. The most comprehensive examination of the global scale of fallowing ended up with nearly 450 million hectares (almost 30 percent of all cropland) fallow at the beginning of the 21st century.[20]

Fallowing is done because of many long-appreciated benefits of the practice: it increases soil's organic matter and its microbial diversity; it reduces soil compaction and improves soil's physical properties and moisture-holding capacity; and by disrupting the life cycles of pests it lowers the need for pesticides.[21] These gains are especially notable in tropical and semi-tropical systems, where fallowing lasts for several years. Fallowing in semi-arid temperate cereal-producing areas (the western US Corn Belt and central North American plains, and the Central Asian steppes) is done in alternate years, in order to store enough water in the soil during the fallow year to support a reasonable yield in the cropping year. Obviously, the universal adoption of double-cropping would affect some of these benefits.

Third, the suggestion that universal double-cropping will be an irresistibly rewarding transformation of today's supposedly less-than-desirable practices ignores the economic realities and individual decision-making.

In many parts of the world dominated by single-crop practices (as is the case in most US regions, where double-cropping is the norm on less than 5 percent of all cultivated land), the adoption of double-cropping would require a farmer to change spring and fall field management, to work against the dominant norm, to incur additional economic costs due to buying seeds, planting, plowing the mature cover crop under, and preparing the field for cash crop planting, and to carry the risk of ending up with lower returns than those producers who follow a standard cropping sequence.[22] Assuming 100 percent cover crop rotation in an academic paper can be done without thinking about the hundreds of millions of farmers who would have to adopt this practice and follow it every year: actually reaching that absolute coverage by voluntary adoption does not appear more likely than instituting it as a legal requirement.

Fourth, agronomists know that many things can go wrong with leguminous cover crops, including poor germination of the cereal crop resulting in patchy crop establishment and lower yields; winterkill where survival was expected; dense roots plugging drainage tile lines; contamination of the next crop; and enhanced release of phytotoxic compounds that suppress the next crop's germination.[23]

Given these realities, it is indefensible to use a single average annual total of 102.8 kilograms of nitrogen per hectare, as did the authors of the universal organic farming study, as the equivalent of nitrogen sequestered by leguminous cover crops and available for the succeeding grain, tuber, or oil crop. To treat the variety of leguminous cover crop species (from alfalfa and clovers to vetches, and from cowpeas to field peas) that would be grown in environments ranging from the tropics to the sub-Arctic and that would be sown annually on the entire area of the world's farmland (1.5 billion hectares) in such a simplistically uniform way is grossly misleading.[24]

Fifth, the universal expansion of cover crops would require an enormous enlargement of the requisite seed production. In temperate regions, cover crop species usually require most of the growing season to mature, and hence growing them for seed necessitates forgoing the cultivation of cereal, oil, or tuber crops on the same land. The first quantitative evaluation of the area needed to supply the United States corn production with cover crop seeds was published in 2020, and it concluded that, depending on the species used, anywhere between about 4 percent and up to nearly 12 percent of the current production area would be required to grow the seeds.[25] And that was only for US corn: extending annual cover-cropping to all cultivated land would require the development and constant operation of a new mass-scale seed industry.

Lessons from the past

If universal annual rotations with leguminous cover crops are highly unlikely, the history of intensive cropping confirms that frequent rotations (once every 2–4 years), as practiced by traditional agricultures, are also not enough to bring about modern high yields.

These rotations, known since antiquity, became common in England and parts of western Europe from the late 18th century, and some of them eventually tripled the nitrogen available to cereal or tuber crops. Remarkably, this innovation was perhaps as significant for Europe's economic development as was steam power.[26] Similarly, Chinese farming depended on them for its high yields for centuries.[27]

But these practices could not ensure the very high annual staple crop yields required by the increasing populations of the 20th century, and they were either augmented or superseded by intensifying applications of inorganic fertilizers. Moreover, while nitrogen is always the nutrient in highest demand, high-yielding crops also require more phosphorus and potassium than provided by indigenous soil resources and recycled crop residues, and these two macronutrients cannot be provided in sufficient quantities by leguminous cover crop rotations.

A load of excrement

Could the only other comparably large source of organic nitrogen—animal manure—be a possible solution? The answer is a resolute no. The best global estimate is that in 2019 livestock voided about 128 million tons of nitrogen, but 70 percent of that was left on pastures, and only about 27 million tons (that is, about 70 percent of nitrogen in manure produced in confinement, stables, and feedlots) was applied to agricultural soils.[28] But in order to displace just half of all synthetic fertilizers (assuming that the other half of the replacement would come from leguminous crops), we would need to recycle at least half of recent annual manure output rather than the around 20 percent we have been applying recently. That means we would have to initiate the mass-scale collection of manure voided on pastures. But even when fresh, these ruminant wastes contain a mere 1–2 percent of nitrogen, and they are dumped in small patches over areas amounting to many millions of square kilometers (the area of the world's pastures is more than 30 million square kilometers!). Just think of the technical modalities and costs of collecting these thinly distributed N-poor wastes and transporting them to regions of high crop productivity.

The alternative—producing more manure in confinement—is an obvious non-starter, because such an expansion would have to be accompanied by an intolerably large increase of crops grown for animal feed rather than for food, and unless those additional animals were pigs and poultry, also by a substantial increase of methane emissions. Another essential consideration: unless highly, and expensively, mechanized, any large-scale expansion of manure application will also call for more labor, a demand running against the long-term trend of declining labor inputs per unit of yields.[29]

My verdict: I find the claim that either of the choices available to organic farming—legume rotations or intensive manure recycling—could supply all the macronutrients needed to produce enough food for the global population "eating as it does today" in 2050 to be—to put it mildly—quite unrealistic.

From annuals to perennials: a difficult path

Completely organic farming would also eliminate all genetically modified organisms (GMOs), all insecticides, fungicides, and herbicides, hoping that strategies to enhance natural defenses will accomplish nearly identical crop protection. Proponents of new, supposedly less disruptive ways of crop cultivation have argued that these natural defenses would be strengthened by adopting a less intrusive approach to cropping through the widespread adoption of staple grain crop perennial cultures.

Having staple crop perennial cultures (permaculture)—planting seeds or seedlings and then, as we do with grasses cut for fresh fodder or hay, harvesting regrown crops indefinitely year after year—would bring a large number of obvious agronomic and environmental benefits, including reduced soil compaction and soil erosion, better water retention, and the reduced use of energy, machinery, and fertilizer. This is not a new idea, and actual development of perennial food (and feed) grains has been going on for decades—although the only major food crop that is now universally recut after the initial harvest is sugarcane. But this tropical grass (a natural perennial species) is not cultivated as a real permaculture. In Brazil, by far the largest producer of this crop, the yield stays good for five or six harvests, and afterwards the field is replanted using new seedlings.[30]

Intermediate wheatgrass (*Thinopyrum intermedium*) has been the most studied perennial grain; it has been widely used in breeding to improve the frost- and drought-resistance of annual wheat, and the leading advocate of permacultures, the Land Institute in Kansas established by Wes Jackson, has been offering it (under the brand name Kernza) for simultaneous forage and grain production, with the grain used in baking.[31] Small seeds and low yields are major drawbacks, but the crop's proponents hope that by the early 2030s seeds should be larger and the plants will be shorter, resulting in a higher yield. In any case, this is where the crop stood in 2022: Kernza was grown on nearly 1,600 hectares in the US; wheat on nearly 14 million hectares.[32] In Russia, a winter-hardy variety of wheatgrass, obtained from Kansas, has been tested for both grain and fodder.[33] Russians have also been developing winter-hardy and drought-resistant perennial wheat (Trititrigia) with high protein and gluten content, again for both grain and fodder.

Perennial rice: how far, how fast?

The development of perennial rice has been made easier thanks to the existence of a perennial wild species of the plant, *Oryza longistaminata*, as well as the fact that ratoon rice (plants regrown from stem nodes after a crop's harvest) may have a similar net energy yield with a lower energy input and production cost.[34] Real rice permaculture got closer with the development of crosses between wild rice and the annual *Oryza sativa*: this work, begun in 1997, eventually led to the introduction of a perennial cultivar in China in 2018.[35]

So far, the most comprehensive report about the advances of perennial rice claims average yields of 6.8 tons per hectare (compared to 6.7 t/ha for annually replanted rice) for eight consecutive harvests over four years, and 2021 plantings amounting to 15,333 hectares (about 0.05 percent of the country's rice-growing area) by smallholder farmers in southern China.[36]

With no need for annual tillage or the transplanting of seedlings, and with less fertilizer and irrigation water, perennial rice reduces material and labor costs. But experimental plantings on hundreds of hectares for three regrowth seasons does not automatically translate to having a truly perennial crop thriving on millions of hectares (China plants annually about 30 million hectares of rice). The major constraints on scaling up the new crop are climate and weeds: in China, the new cultivar cannot survive winter in latitudes above 26°N and there is no annual tillage to suppress weeds. While Chinese research on perennial rice continues, one of the participants in this effort concluded that (as of 2022) it is hard to imagine that the new cultivar will replace ordinary annual crops in a large part of China's rice-growing area.[37]

The development of cereal permacultures (or at least multi-year crops) will certainly continue, but it is most unlikely that within two or three decades these new cultivars will be supplying an important share (not even a fifth) of global staple food needs.[38]

Perennial polycultures

Even less likely is the appearance of viable perennial polycultures with several different crops grown simultaneously in the same field. A step toward this option is the ancient practice of planting annual maslins—mixtures of cereal species (wheat and barley; wheat and rye) that were not uncommon in the past in parts of Europe, Asia, and North Africa. The desirability of these mixtures has been recently promoted as a step toward sustainable agriculture.[39] Robert Loomis, an American plant scientist, dispelled the notions regarding both perennial grains and polycultures nearly two decades ago (in a thorough critique published posthumously only in 2022), by stressing the challenges of increased water needs and agronomic management, and the inescapable compromise between perenniality and yield. His conclusions, summed up in an exemplary way, still stand: "If yield is to be maintained, external inputs are essential, regardless of life-habit. Polycultures of perennial grains are seen to have little potential for producing sufficient food to serve as alternatives for current production systems."[40]

Genetically modified ideas (and realities)

And taking one more step along the progression of improbability is to believe in the imminent commercial availability of genetically redesigned food crops that would be able either to convert solar radiation with substantially higher efficiency or to enter into symbiosis with nitrogen-fixing bacteria and provide most of the needed nitrogen. There are two basic options to get staple cereals to produce their own nitrogen: turn them into hosts of symbiotic Rhizobium bacteria (that is, make them legume-like in this respect) or through genetic transformation, by encoding symbiosis as a permanent trait by introducing nitrogen-fixing (*nif*) genes directly into the genome of cereal plants.[41]

Work on nitrogen-fixing cereals has been going on since the 1970s, but reports of progress have remained strictly within basic science and laboratory stages. There is no *nif*-containing wheat growing in experimental fields, and even if the nitrogenase encoding in staple cereals succeeded, we would have to make sure that it did not have a major negative impact on yields or on other desirable plant traits. And then, as with all transgenic crops, we would have to overcome the reluctance to adopt them: the US and Canada may abound in transgenic corn and rapeseed, but these cultivars are forbidden within the EU. Responsible researchers who have studied this problem for decades know that we cannot give an answer to the question of how long it will take to have nitrogen-fixing cereals.[42]

Genetically engineered crops that could grow faster and use fewer resources while raising the maximum conversion efficiency of photosynthesis would make an even greater contribution to future food production. Perhaps the most promising option is to improve the operation of RuBP carboxylase-oxygenase or Rubisco, the photosynthesis-enabling enzyme, by endowing it with a CO_2-concentrating mechanism (making it more like a C_4 photosynthetic pathway); this would promote the enzyme's yield-increasing activity while minimizing its undesirable, yield-lowering oxygenase activity.[43] These efforts have been underway for decades, accompanied (as is often the case in the early stages of innovation) by unrealistic expectations. In 1986, plant geneticist Chris Somerville concluded that "recent advances in the development of techniques for the manipulation of gene structure *in vitro* and genetic transformation of plants have brought the goal of directed genetic modification of RuBP carboxylase-oxygenase (Rubisco) within grasp."[44] But nearly four decades later, that goal remains elusive.

A 2020 review of the prospects for engineering biophysical CO_2-concentrating mechanisms (CCMs) into crops to enhance their yields offers a more realistic conclusion: after decades of research, "efforts to reconstitute these CCMs into land plants will take many years but are already yielding encouraging preliminary results."[45] How encouraging is still a matter of debate. In 2022, a group of Illinois researchers claimed that they had found a bioengineered solution to the dissipation of sunlight as heat by crop leaves (a common problem that reduces crop photosynthesis), resulting in up to 33 percent higher yields during brief field trials.[46] But this claim was quickly contested by another group of plant scientists, who pointed out the questionable premises for the soybeans study and the invalid field protocol to evaluate the results, concluding:

"We still see no evidence that crop yield can be substantially increased by alteration of crop photosynthesis."[47]

Of course, the slowness of these advances in the development of crops capable of either fixing their own nitrogen needs or performing with higher photosynthetic efficiency is not surprising; improving on fundamental limits imposed by hundreds of million years of plant evolution is an extraordinarily difficult challenge, with meager prospects for any early commercial breakthroughs and for the world's harvest being transformed by new super-efficient staple crops.

Cultured meat

While there are no prospects for any immediate commercial deployments of nitrogen-fixing cereals or C_3 crops with enhanced carboxylase activity, there have been many reports of highly promising advances in producing cultured, or *in vitro*, meat. The first cultured meat patent was awarded in 1999; the first small piece of lab-cultured beef was produced (at an exorbitant cost) in 2013 in the Netherlands; the first cultured meat company was launched in 2016; and the first small-scale sales of chicken nuggets grown in a bioreactor took place (after two years of regulatory process) in a Singapore private members' club in December 2020.[48] By 2021 the total investment in the new industry reached nearly $2 billion, and JBS, a Brazilian company that is the world's largest meat processor, invested $100 million and announced that it intended to start selling cultured meat by 2024.[49] Moreover, these innovative developments now also include research to produce cultured fish (both freshwater and marine species) and shrimp meat, as well as eggs and milk.[50]

Are we really close to the routine, mass-scale production of animal foods without animals, and getting steaks, pork chops, and chicken breasts without slaughter—or is the imminent arrival of the cultured animal product industry just another case of vastly

No small aspirations for cultured meat: from a Petri dish to feeding the world.

exaggerated expectations? As you might expect having read most of a book that favors skepticism over dubious claims, a more systematic consideration is revealing.

How does it work?

Cultured meat production is a complex, multi-step process that begins with biopsied adult skeletal muscle stem cells taken from donor animal tissue.[51] These cells proliferate in cultures designed to resemble the conditions inside animal bodies: first in small flasks filled with oxygen-rich growth media composed of aqueous solutions of glucose, amino acids, growth factors, vitamins, and salts.

Decomposed (hydrolyzed) yeast, soy, rice, and microbial substances might be added to these basal media solutions to provide additional sources of the compounds needed to produce myoblasts (embryonic precursors of muscle cells). Growing cells are subsequently transferred to a series of increasingly larger bioreactors. In the process, a tiny mass of myosatellite cells taken from animal tissue turns into proliferative myoblasts, and these, after immersion in a modified culture medium, differentiate to non-proliferative muscle, fat, and connective tissues that fuse into myotubes (multinucleated fibers), with the entire process taking (depending on the species) two to eight weeks. Scaffolding that provides structural support for cells to grow and to produce meat-resembling tissue must be edible or dissolve before consumption.[52] Nutritionally, this cultured product is a perfect equivalent of lean meat (much like a chicken nugget) but it lacks the structure, mouthfeel, and fat content of cooked beef or a pork cut.

The mass-scale production of cultured meat takes place in bioreactors (large stainless-steel containers) under carefully controlled conditions. The most common type is a continuous stirred tank reactor in which cells grow in suspension. The biopharmaceutical industry has used bioreactors with volumes of up to 20,000 liters, but these are still too small for mass-scale cultured meat production. But no matter what size, successful bioreactor operation can proceed only within tightly defined parameters that require the constant and sophisticated monitoring of temperature (optimal at 37°C for mammalian cells), acidity (pH 7.4, with a variance of 0.4), dissolved oxygen and carbon dioxide levels, glucose and cell concentrations. Reactions must proceed in a sterile environment, and this requires the elimination of any bacterial, fungal, and viral infections, as well as of cell cross-contamination, and the presence of antibiotics; maintaining these conditions needs various methods of sterilization, including flash pasteurization and microfiltration. Not surprisingly, these conditions and processes require considerable energy inputs for heating, agitation, sterilization, cleaning, and sanitation.

Making meat, making money

In March 2021 the Good Food Institute, an organization of alternative protein producers, published a techno-economic analysis that predicted the cost of cultured meat will decline from more than

From pastures and CAFOs to sterile steel: will a significant share of the world's meat come from bioreactors?

$10,000 a pound to just $2.50 a pound by 2030, a stunning 4,000-fold cut that would make such meat competitive with the natural product![53] This claim led Paul Wood, a former executive at Pfizer Animal Health, to commission an independent analysis of the report. It dismissed the institute's claim for the following reasons: the finished product for consumption remains undefined; the estimated cost excludes purification, packaging, and distribution; cultured cell growth will demand permanent sterile cultures at pharmaceutical levels that are not feasible at a food production scale; cell-based protein will have to be enriched by necessary vitamins and minerals and made flavorful by the addition of oils and fats.

Perhaps most fundamentally, given that the pharmaceutical industry has improved the productivity of its cell-based medicinal products between 10- and 20-fold in the past 15–20 years, it is unlikely that cell-based meat production could be cut to less than 1/1,000th of their current cost. Huw Hughes put the real cost of 1 kilogram of cell culture in excess of ~$8,500–$36,000, compared to the wholesale price of trimmed US chicken meat at $3.11 per kilogram in February 2021. Moreover, "these cost estimates should be revised upwards as the consumer product is defined, including the methods for purifying, processing and packaging the raw ingredient into a tasty product that is free from harmful contamination."[54]

This was the second analysis strongly disagreeing with the Good Food Institute's optimistic conclusion. In December 2020, another step-by-step evaluation of the *in vitro* meat process found a series of technical challenges (including cell metabolism, reactor design, cost of ingredients, and cost of high-throughput facilities) that make it unlikely for the culturing process to succeed at food scale—that is, producing millions of tons of product a year.[55]

Scale

As for the necessary scaling-up of the bioreactors, a few basic comparisons with existing processes illustrate the daunting magnitudes required for commercial production at rates that would displace significant shares of natural meat. The model facility in the Good Food Institute's analysis would produce 10,000 tons of cultured meat a year, and its largest stir-tank reactor would hold 10,000 liters of proliferation cells before their transfer to smaller 2,000-liter reactors for differentiation and maturation. Its total bioreactor volume would have to equal nearly a third of the combined volume now operated by the world's entire biopharmaceutical industry (about 63 million liters).[56] This means that it would take 300 of these facilities to produce the equivalent of just 1 percent of today's global meat output.

And because the pharmaceutical industry, and particularly the large-scale batch preparation of vaccines, is the closest functional equivalent to what would have to become a new kind of industrial production, it also provides relevant comparisons of required energy inputs. So far, the most detailed attempt at an anticipatory life cycle analysis of *in vitro* biomass cultivation of cultured meat in the US concluded that the practice would, as expected, require lower agricultural inputs and less land than livestock, but that this would come at the expense of more intensive energy use. In fact, the global warming potential of cultured meat would likely be larger than for producing pork and poultry![57] This finding appears less surprising when one considers the result of one of the most surprising analyses of carbon footprint comparing the global pharmaceutical and car industries: when using the same analytical model and the same methodology, the former has about 55 percent higher intensity (when measured per million dollars) than the latter.[58]

And even if expected advances in cultured meat production were to bring significant energy savings, the challenge of scaling up an entirely new kind of industry would remain. The global production of meat is now more than 300 million tons a year; if *in vitro* meat were to supply just 10 percent of that mass, we would need to grow more than 30 million tons of it every year. Here is another comparison with the pharmaceutical industry: global sales of animal antimicrobials (fed to chicken, cattle, and pigs) have been close to 100,000 tons a year, and because antimicrobial consumption in animals is about twice that by humans, the global total of antibiotic production in 2020 was on the order of 150,000 tons a year.[59] Developing a new global cultured meat industry supplying 30 million tons a year would thus require a brand-new industrialization endeavor 200 times larger than the world's preparation of antibiotics, a kindred industry developed during the past three-quarters of a century.

If you believe the marketing

None of this seems to be much of an obstacle for cultured meat proponents. The Good Food Institute's report on the state of the industry in 2021 highlighted record new investments, the arrival of new companies (a total of 107 in 2021, nearly double the 2015 total), and concluded that the pilot scale phase (2019–2022) is about to end as the demonstration scale (producing thousands of tons of cultured meat a year) was to begin in 2022—followed by the industrial scale, with millions of tons produced every year. But the update to the report did not go much beyond writing about "an industry on the cusp of transforming 12,000-year-old ways of making meat."[60] In January 2022 a report by ResearchAndMarkets.com projected what must be termed a rapid demise of real meat, foreseeing that by 2040, 60 percent of all meat—by that time, that would be equivalent to about 250 million tons of tissue worldwide—will be grown from cells inside bioreactors.[61] And in April 2022 the Dutch government awarded a €60 million grant to Cellulaire Agricultuur Nederland, an association of universities, companies, and consultancies, in order to build "a fully fledged cellular agriculture ecosystem" that would maintain the country's position as the EU's largest meat exporter while eliminating its nitrogen pollution problems.[62]

That would be a stunningly unprecedented transformation of food production, ranking in importance right along with the domestication of animals that began some 10 millennia ago. But, as with any projections concerning an industry that still has not seen even a few years of viable small-scale commercial operation, caution should come first. The Good Food Institute's report for 2022 showed new "cultivated meat" (now the preferred term) companies, and it listed a number of product "prototypes" (ranging from chicken and burger to schnitzel and fish balls)—but it had no news of the promised mass market production, and even in 2023 there was no *in vitro* meat on any supermarket shelves.[63] And even plant-based "meat" is not surging: in 2022 its US retail unit sales fell by 8 percent.[64]

Readers who will be around in 2040 will see who was right. Will it be the people with long experience in the biopharmaceutical industry (whose arguments I find persuasive), who think that, for a variety of reasons, growing cells in bioreactors cannot be readily scaled up to constant deliveries amounting to tens of millions of tons of cultured meat a year—and hence that we will not see the rapid transmutation of billions of chickens, about a billion heads of cattle, and hundreds of millions of pigs into a mass of stainless-steel bioreactors loaded with myocytes? Or will it be today's enthusiastic cultured meat proponents and investors, who see substantial commercial deliveries taking place before 2030 and who believe that not only meat animals will be in full retreat in less than two decades but also that cultured fish and crustaceans will do away with both wild catches and aquaculture and that eggs and milk will, too, come from bioreactors?

If you believe the best evidence

Given a worrying amount of misinformation, I thought it necessary to include these assessments of questionable options and to explain why I do not think that they will make substantial (and even less likely, fundamental) differences in securing adequate food supply by the middle of the 21st century. I have made my point by referring to historical precedents (fundamental transformations take time; the simplicity of early promises turns out to be far more complicated) and by pragmatic concerns; by advancing and perfecting known and effective solutions rather than hoping for rapid and unprecedented gains of new—and too often uncritically appraised—approaches. Of course, I readily admit that on some accounts I might be proven wrong and that major advances could turn some of today's dubious options into widely accepted commercial realities. That would be a great outcome, and a welcome development whose impact would not be diminished by advocating less stunning but in the end effective solutions.

As I have said in previous books, I am not a pessimist or an optimist, I am a scientist. Of course, scientific consensus changes as new discoveries and new assessments are made, but the prudent, and properly skeptical, attitude during the 2020s is neither to endorse any mass-scale adoption of veganism nor to invest high hopes in the imminent large-scale production of affordable cultured meat. We know that symbiotic interactions of plant and animal foods improve human health and that avoiding animal foods would pose greater risks of nutritional deficiency for large shares of the population, especially for children, the elderly, and nursing mothers—and that many people would not thrive even if vegan diets were carefully designed to take care of the low content and low bioavailability of some key micronutrients.[65] And we know that too many promising and supposedly transformative innovations have not ended that way.[66] Ambitions and aspirations are one thing, realities another.

8. Feeding a Growing Population: What Would Work

I would be glad if some of the skepticism concerning major breakthroughs in food production proved to be wrong, and if we had major early advances in photosynthetically more efficient crops or in mass-scale and affordable cultured meat—the two breakthroughs that would make it easier to produce more food with reduced environmental impacts. But it is encouraging to think that we could succeed in securing food for more than 9 billion people without such profound changes, by relying on well-proven solutions and on their likely future gradual improvements. Moreover, this remarkable accomplishment should also be accompanied by reduced environmental impacts.

A brief recap of some of the issues explored in this book explains the magnitude and intricacies of the combined challenge of producing higher output with moderated inputs and lessened undesirable effects. The expansion of new arable or pasture land in the tropics leads to additional deforestation; widespread and repeated planting of row crops subjects soils to greater erosion (before their canopies close and protect the ground against the direct impact of rain); more frequent monocropping (corn after corn after corn) and more common cultivation of a limited number of crop species reduces the cropping system's diversity and provides more opportunities for established pests and weeds.

Effects on soils range from always present wind and water erosion; soil compaction (by repeated use of heavy machinery); loss of soil organic matter; soil salinization (due to excessive irrigation in arid regions); and soil contamination by heavy metals (from fertilizers and organic and industrial wastes). Interventions in the water cycle include rising shares of water use claimed by irrigation;

falling water tables of numerous aquifers that have been subjected to excessive pumping; contamination with soluble compounds (above all, nitrates from intensive fertilization); and the resulting recurrence of coastal dead zones. As already noted, food production is a major contributor of greenhouses gases: CO_2 from land use changes and the decrease of organic matter; methane from anaerobic fermentation in anoxic wet (rice) fields and in the stomachs of ruminant animals; and nitrous oxide, a product of the denitrification of nitrogenous fertilizers.

Fortunately, in many cases, simple shifts in dominant practices or single new choices will have numerous concurrent effects and benefits. Crop rotations can improve soil quality (as more organic matter and more bacterially fixed nitrogen get plowed in after a season of a leguminous cover crop); substitution with a more water-efficient crop can reduce the demand for well-water withdrawals; choosing to eat more chicken meat than beef, a shift that has been underway for decades in most Western countries, has a large cascade of environment-sparing consequences because chicken needs only a fraction of the feed required to produce the same mass of protein (and hence a fraction of cultivated land, water, and fertilizer). There is no shortage of entries in the "what will work" category, and their detailed consideration would be a great topic for another book. But here are some of the measures that can be widely adopted.

What will work

When dealing with excessive environmental burdens—as we certainly are—greater efficiency and the voluntary adoption of a lower intensity of inputs may not suffice, and there will be the need for restrictions on extremely intensive food production. These limits would need to be based on carefully established environmental capacities, and the already noted Dutch limit on nitrogen loading per unit area is an excellent example of a measure designed to reconcile high productivity with environmental limits.[1] Better soil management, aimed at increasing soil's organic content and water-holding capacity, should be a constant concern, and these efforts should be in even greater favor due to their carbon-sequestering capabilities: soil is the world's largest reservoir of organic carbon, and hence even relatively small carbon-storage gains translate into meaningful absolute removals of CO_2 from the atmosphere.[2]

Where possible and practical, multicropping (with its many beneficial effects that outweigh some inevitable, and earlier discussed, downsides) should be a frequent or dominant choice, and if economically acceptable, properly managed recycling of animal manures (the key component of organic farming) should be encouraged. Continued and gradual yield improvements can be expected from the breeding of major cultivars, and major productivity gains would follow the introduction of crops able to cope better with specific environmental limits. For example, the development of rice cultivars that can tolerate water flooding and drought would be especially welcome, as would the introduction of drought-tolerant wheats with improved ability to survive more frequent droughts in our warming world.[3]

High-tech for crops

So-called precision agriculture has developed from an intriguing idea to an increasingly affordable and, in both North America and Europe, widely used tool to enhance productivity while minimizing inputs and environmental impacts.[4] Several distinct techniques fall under this new type of farming. Soil monitoring identifies differences in nutrient and water content, and global positioning systems (GPS) guide machinery to treat fields (fertilize, apply herbicides and insecticides, irrigate) with discrimination and accuracy. GPS can also guide machines across fields in the most efficient patterns, and lasers can be used to level fields (resulting in more efficient irrigation and reduced runoff). Many farming variables, once trickier and more expensive to monitor, can be now collected less expensively by relying on remote sensors carried by low-flying drones.

Further adoption of tech-powered farming would do a world of good, especially if taken up by smallholder farmers in Africa, which would require expanded "extensions services"—such as soil testing and accurate agronomic advice (practices long taken for granted in high-income countries)—and more affordable options. If most smallholders around the world were able to benefit from such knowledge, productivity might be improved by such simple measures as the better matching of crops and soils, more rewarding crop rotations, and the choice of more pest-resistant varieties.[5]

In combination, these measures—the further improvement and extension of the best farming practices—pursued across decades would bring about a substantial transformation of modern intensive agriculture, and interested readers can find a wealth of publications detailing these practices, their recent advances, and their possible future contributions. But in this book's closing chapter, I want to concentrate on two potentially highly effective changes, two strategies that involve *doing less* rather than *doing more*, and whose adoption would help to produce global food needs by the mid-21st century with less impact on natural ecosystems and with lower material and energy inputs—a win/win scenario.

Globally, this means looking to reduce the world's enormous food waste (a concern that is finally receiving closer attention), and in affluent countries it means steps toward moderating high rates of meat consumption and changing its composition (these processes have been underway for some time, but they could be advanced still further). These measures amount to relying on ordinary incremental changes, devoid of any stunningly disruptive innovations and radically transformative upheavals; that's to say, these changes are possible and necessary but are often ignored by the media and a number of writers of popular non-fiction who instead focus on the unrealistic. As I will explain, these measures— deemed unnewsworthy and unexciting—pursued systematically and over the long term would bring substantial—newsworthy and attention-grabbing—results.[6]

Food waste

I put food waste first because its reduction represents the most obvious—yet chronically most neglected—way to expand food supply without any additional inputs. After all, the wasted food has been already harvested and most of it has been also processed and distributed, and hence any reduction of this waste represents a nearly free (or very low-cost) opportunity to expand the available food supply while reducing the environmental impacts associated with food production. The cumulative effect of these reductions is substantial: in the US, food that does not get eaten requires a quarter of all water used in field irrigation, it embodies at least 4 percent of the country's crude oil consumption, and it burdens landfills with more than 35 million tons of waste a year.[7]

Reasons for food waste range from obvious overproduction (especially when considering aging populations of all but a few high-income countries, with their lower average daily demand) and the relatively low cost of food, to the decline of home cooking and adherence to best-before dates. American consumers cite their concerns about food poisoning as the most important reason for discarding food. But if stored in a refrigerator many products are perfectly safe to eat after their expiration dates: for example, unopened yogurt or buttermilk a week (even two weeks) later, eggs three to five weeks later, but fresh meat cuts no more than 3–5 days later.[8]

Of course, some waste is unavoidable (peelings, bones, eggshells, tea leaves) and some is avoidable but excusable—a part of the normal inefficiencies of human lives. But most food waste is unnecessary and avoidable, and yet until recently hardly any effort went into quantifying it: for decades the preoccupation of national and global food-related institutions was with maximizing food outputs, not with minimizing post-harvest food losses. Shifting this emphasis is not easy, and a long road lies ahead for any food-waste reduction efforts. I have no illusions about the speed and the extent of taking advantage of opportunities that, in efficiency parlance, constitute low-hanging fruit.

How we could do it

The most apposite analogies are reinsulating older buildings to meet modern efficiency standards and recycling municipal waste. These tasks are not always easy, but they are readily doable and they yield long-term rewards. The benefits of upgraded housing last for decades and generations—yet despite appeals, promotions, and rebates, retrofitting insulation remains a niche activity. Just compare the investments flowing into producing more energy (solar, wind) or introducing new energy converters (electric cars) with those going into triple-pane windows and fiberglass insulation.[9] Similarly, recycling some materials is far more difficult than reusing others (plastic packaging vs. aluminum cans is perhaps the best example of this contrast), but even imperfect, partial recycling beats the mindless discarding of everything.[10] Neither is surprising: investments in promising new techniques, the construction of new projects, and the promotion of new approaches have always had greater appeal and more institutional and financial support than the unglamorous but effective improvement of the old ways and the reduction of waste.

As I have tried to show in this book, every solution must start with a good understanding of the problem's magnitude and complexity, and the challenge of reducing the indefensibly high level of food waste has been, finally, receiving more attention. Even the FAO, whose previous concern was almost solely with food production, finally contracted the Swedish Institute for Food and Biotechnology to produce the first global food waste study.

Published in 2011, it put the shares of global food loss and waste at about 30 percent for cereals, 40–50 percent for root crops, fruits, and vegetables, 20 percent for oilseeds, meat, and dairy foods, and 35 percent for fish.[11] The highest wastage was, predictably, in the EU and North America (on the order of 100 kilograms a year per capita), while the losses in sub-Saharan Africa and Southeast Asia were an order of magnitude lower.

This was followed in 2014 by the FAO's Global Initiative on Food Loss and Waste Reduction, but we have no subsequent global appraisals.[12] However, the UK's WRAP studies, launched in 2007, have documented some progress in the country's waste reduction.[13] The initial study tried to quantify several fundamental variables, and it concluded that UK households waste 6.7 million tons every year, or about a third of what they purchase; that nearly 90 percent is collected as municipal waste and mostly landfilled (contributing to methane generation); that at least three-fifths of this waste was avoidable; and that the unavoidable food waste (peelings, meat carcasses, teabags) added up to less than a fifth of the wasted mass.

Potatoes led the list of wasted foods, followed by sliced bread, apples, meat, and fish, and about half of all food thrown away was fresh. Perhaps most remarkably, more than a quarter of all avoidable waste was thrown away whole or unopened. In 2012, the second WRAP study reported a 15 percent reduction (1.3 million tons) of household food and drink waste since 2007, even as the number of households increased by 4 percent; this improvement led to an 18 percent reduction in food waste collected by municipal garbage services.[14] Changes in packaging and a better redistribution of food were among the main changes making a difference.

Also in 2012, the US Natural Resources Defense Council published its analysis *How America Is Losing Up to 40 Percent of Its Food from Farm to Fork to Landfill*.[15] And certainly the most unexpected was the Canadian 2019 assessment claiming that 58 percent of all food produced is "lost or wasted," and that of this total 32 percent "could be rescued to support communities across Canada."[16] These institutional reports have been the ones that have gotten the greatest public attention, but food loss has seen a rising wave of academic studies and management analyses.

Papers listed on PubMed tell the global story, with publications on food waste being an uncommon topic (fewer than 50 new items a year) for decades, rising to 139 in 2000, 753 in 2010 (more than fivefold in a decade), and 3,021 in 2020 (a fourfold rise during the 2010s).[17] Encouragingly, and as befits the complicated nature of the problem, this new flood of publications has gone beyond generalities, as it addresses methodological matters and all major sources of food waste, and looks for many specific solutions. Analytical frameworks are important to avoid misleading comparisons and excessively large estimates: some studies define waste as both inedible and edible components of food, while others limit themselves to edible parts; some studies consider all stages of the supply chain, others focus on processing or household waste.[18]

Practical guidance

Food waste studies should be also sufficiently specific in order to provide practical guides for our actions, above all by concentrating first on the practices and habits that could bring the greatest rewards. A recent American study, based on nationally representative food intake data from 16 years (2000–2016) of the National Health and Nutrition Examination Survey (and linked with published data on food waste and food prices), is a good example of such practical guidance. The study concluded that 27 percent of daily average per capita food expenditure was wasted, 14 percent was inedible, and 59 percent was consumed—and it identified the two greatest daily food waste expenditures: meat and seafood consumed outside the home (adding up to about 7 percent of daily food expenditure), and fruits and vegetables purchased for consumption at home (about 5 percent of daily spending on food).[19]

A recent global study on food waste concluded that consumers discard much more food than is widely believed; in fact, the most widely cited global estimate of food waste may be underestimated by a factor greater than two—the actual rate being about 540 kilocalories a day per capita rather than only about 215 kilocalories.[20] Moreover, the expected large national differences appear to follow a linear-log relationship with consumer affluence: food waste starts to emerge when daily expenditures rise above $6.70 a day per capita. First, it rises rapidly just above that level (increasing with increasing affluence), but soon the rate of increase tapers off. Another analysis confirmed the overall rising-and-saturating trend, but it showed a large variability, with some low-income countries wasting as much as the richest nations.[21]

But this does not invalidate the desirability of starting with interventions designed to prevent further growth of food waste before consumer incomes begin to rise. Hardly any such effort took place in China as incomes multiplied and as food supply rose from the subsistence levels of the late 1970s to the average per capita rate that has been surpassing 3,000 kilocalories a day. Although data from China's Health and Nutritional Survey indicate that household food waste declined by about 20 percent between 1991 and 2009, higher incomes have allowed for more frequent eating out, with more waste generated in eateries.[22] In China there is another factor of great importance that contributes to the nationwide wasting of food: the need to preserve *mianzi*, to save face. This means being sensitive about one's reputation and prestige and avoiding shame, and where wealth and hospitality are concerned this attitude is displayed by the overconsumption of food on social occasions, with hosts ordering more food than can be eaten, and with the share of waste increasing with the size of banquets.

This well-known traditional behavior has been confirmed by recent empirical studies that established a strong link between a person's vanity score and food waste, particularly when eating outside the home.[23] In contrast, a new study showed still very low rates of household food waste in the country's rural areas, between just 1.1 percent in the poorest and 2 percent in the richest region.[24]

Whatever the actual total mass or rate of China's food waste, it has reached levels that have become a matter of public concern, and led to the adoption of a new state law in April 2021. This is the world's first legislative measure of its kind, with the state promising to take "technically feasible and economically reasonable measures to prevent and reduce food waste" and calling for "socially responsible, healthy, resource-saving, and environmentally friendly ways of consumer spending" while advocating "a simple, moderate, eco-friendly and low-carbon lifestyle."[25]

Preventive measures

As is the case with material recycling (with disparate categories ranging from kitchen waste to the reinforced concrete from demolished buildings and roads), preventing food waste and reducing food losses must include a multitude of different efforts, some proceeding on industrial scales in a limited number of large facilities (commercial fresh and frozen food storages), and others involving many individual and everyday actions such as managing the contents of household refrigerators. Logistics solutions for reducing food waste include better forecasting of demand, maintaining the right levels of stocks, price reductions, reviews of product varieties, and improvements in packaging design (more on that later).[26] Nearly all high-income countries should lower their overall food supply; even with 20 percent aggregate loss, anything above 3,000 kilocalories a day per capita leads to higher waste, and yet all but two EU countries (Bulgaria and Slovakia) have been recently above that level, with a third of them being above 3,500 kilocalories a day, the same as the US and Canada.[27]

All countries need to minimize wholesale storage and distribution losses. This has many technical and managerial solutions: large storage facilities can trace electronically every shipment's quality and projected durability and be equipped with temperature, humidity, and other sensors to maintain optimal storage conditions, and modern distribution methods must be gradually adopted in lower-income modernizing countries, but the costs of these measures are significant.[28] Proper storage and handling as well as flexible pricing and prearranged food donations can make a difference at retail level.

What can we do?

At the household level, efforts must start with awareness of the problem and with the understanding of realities that motivate people to minimize their waste. Only about a quarter of participants in an American study claimed to be very knowledgeable about reducing food waste, and they cited saving money as the leading motivation, followed by (almost equally important) setting an example for children by managing the household efficiently and thinking about hungry people. Reducing energy and water use and greenhouse gas emissions was the least important consideration.[29]

Household food waste can be reduced by such simple adjustments as modifying the size or kind of packaging to increase the likelihood that food will be consumed before it becomes waste, and to better preserve the contents.[30] One of the most high-tech options is "intelligent" refrigerators, which remind users about the state of the food inside, but given their cost vs. perceived utility, they will not be a common sight anytime soon.

As for eating out, the obvious first step in North America would be to reduce the portions served, a change that would not result in any starvation servings but that would merely bring them closer to the more sensible amounts offered in European and Asian eateries.[31] In many restaurants where guests commonly choose to dispose of their food leftovers, placing a small card reminding the customers of the magnitude of prevailing food waste, as well as the majority's expectations (most people want to see food waste reduced), can result in a substantial increase of requests to take away the leftovers.[32] Even modest food waste reductions would translate into considerable cumulative savings.

Systemic issues

The most significant result would be obtained by not putting excessive food supply on the retail market. About 900 million people are living in countries with an average per capita food supply above 3,300 kilocalories a day. There is no conceivable defence of this rate: we know from actual dietary intake studies that the typical per capita adult food intake in high-income economies (with their largely sedentary or, in any case, not physically demanding occupations, and with their rising mean age) are on the order of 2,000–2,200 kilocalories a day. Hence even a 30 percent food waste rate would raise the needed supply to no more than 2,900–3,100 kilocalories. Consequently, even if affluent countries remained relatively as wasteful as they are today but simply reduced their food supply to an average of "just" 3,000 kilocalories a day, that would obviate the equivalent of food production sufficient for some 130 million people (roughly the combined population of France and Italy) eating at the same level, with the attendant prevention of food waste and with the commensurate easing of environmental burdens.[33]

Food waste could also be lessened by reducing consumer choice. This does not imply any drastic cuts in selection; one can be firmly in favor of food variety (both as an essential component of nutritional quality and as a precondition for diverse ways of cooking), but there is surely no need for American supermarkets to carry 40,000–50,000 different items of food, especially not given the fact that as recently as the late 1990s they stocked "only" about 7,000 items and were not beset by incessant complaints about poor choice.[34] Inevitably, the recent super-proliferation of variety must be accompanied by greater waste.

An obvious place to start is to have a slightly reduced variety of those items whose production requires relatively large environmental footprints: yogurt is one of the best examples. In the US of the late 1960s yogurt was a marginal product sold in limited variety; by 2020 a typical US supermarket carried some 300 different kinds, ranging from old-fashioned plain variety (though that, too, comes with different fat levels) to scores of Greek (typically high-density, high-fat) yogurts, and from Icelandic skyr flavored with cherries (as if such a combination were possible in Iceland!) to vegan and "probiotic" kinds.[35] In fact it may have been this overwhelming variety that has led to the recent decline in American yogurt sales, as consumers are simply overwhelmed by choice.

An unpopular option

I have left the least popular option for reducing food waste for last: despite recent perceptions of high expense in light of war-fueled price rises, food in high-income countries has been recently cheaper than at any time in history, and while the argument for more expensive food may be politically inadmissible, in a book more interested in science than politics, it deserves a critical appraisal. In 2020 the share of the average American family's disposable income spent on food reached the lowest point ever, just 8.6 percent, compared to 25 percent in 1930 and 15 percent in 1965, with the EU averaging about 12 percent, Japan 26 percent, and China nearly 30 percent (but that includes also alcoholic beverages and tobacco).[36] The Covid pandemic and Russia's invasion of Ukraine began to reverse the long-term decline of this spending, but we will have to see how strong and how long-lasting this effect might be.[37]

There is extensive economic literature on the price elasticity of demand (the relative impact on demand given a change in price, which is expectedly low for staples and relatively high for luxury foods), but as food waste takes place across the entire consumption range, higher prices (with their inevitable effect on the poorest segments of populations) would not be the best way to regulate waste.[38] In contrast, the argument that we are not paying enough for food in order to reflect more closely the real environmental costs of its production is irrefutable—but that argument is part of a much larger problem of avoided external costs (true for energy and all materials) rather than a convincing reason for lowering waste through higher prices.

Eating less of more environmentally friendly meat

The second rewarding "doing less" approach to lowering agriculture's impact on the environment in high-income countries—reducing average meat intakes and changing its composition—works with obvious multiplier effects. Given the inescapably low efficiencies of converting plant feeds into animal foods (see this book's fourth chapter), every unit of reduced meat consumption means eliminating the need for, mostly, 2–5 units of feed when not producing chicken or pork, and well over 10 units of feed when not producing a unit of beef or mutton; this, in turn, has obvious repercussions for reduced applications of agrochemicals, less irrigation, and fewer field operations (reduced diesel use, reduced soil compaction).

And this approach has a far higher chance to make a real difference than the two now so often extolled options. As already noted, veganism will not be universally accepted. Vegetarianism could (some would say should) become a much more common choice, but if it includes significant shares of fruits and nuts then it may not result in dramatic energy and water savings compared to the diets it would be replacing. Cultured meat may remain of marginal (rather than of decisive) importance for a long time (see the previous chapter), and plant-based meat substitutes are unlikely to capture large shares of the market (see the sixth chapter). In contrast, moderating today's high rate of animal food consumption in affluent countries, and changing its composition, is both rational and highly desirable. Above all, it is eminently doable, because it can be achieved without affecting the adequate supply of healthy nutrition and without posing sudden economic losses for producers.

Indeed, in several countries it has been done already, in some cases to a surprisingly large degree, and hence the pursuit of this adjustment would mean nothing more than maintaining, and perhaps accelerating, longstanding consumption trends. In other countries it would take additional measures to accelerate the gradual retreat and to pursue a deliberate, explicit strategy that can bring substantial cuts in average intakes and hence notable reductions in demand for feed and in the total number of large centralized feeding operations, and hence lessen agriculture's overall environmental footprint—all without resorting to bans and rapid dislocations.

Success stories

Notable examples of decreasing overall meat consumption in nations known for their historically high rates of carnivory are Denmark (25 percent down since the peak level in 1992) and Germany (17 percent down since the mid-1990s), and the rates of these declines were matched or surpassed in the US, where red meat consumption has declined by 25 percent since its peak in 1971 and the demand for beef, environmentally the most demanding meat, has been reduced by 37 percent since its peak in 1977.[39] These shifts have been driven by other factors besides price differentials: changing tastes, ease of preparation, and, of course, health and environmental concerns.

But these rates have been derived from food balance sheets, and in some countries actual food consumption data obtained from periodic household surveys indicate even lower intakes. If we take the French (not FAO's) food balance sheet, we see that in 2020 average per capita consumption of red meat (*viande de boucherie*: beef, veal, pork, lamb, horse) was about 65 kilograms of carcass weight a year, and that would be 1.25 kilograms a week, equivalent to some 550 grams of retail cuts. Yet actual consumption surveys indicate that the majority of French people are already "small consumers" (*petits consommateurs*), eating weekly less than 500 grams of red meat, the limit recommended by the National Health Nutrition Program.[40]

On the road to moderation: French meat consumption

As expected, there are disparities among genders (adult men consume about twice as much meat as adult women) and income groups, but people consuming at the lowest rate (less than 100 grams a week) were more numerous than people eating more than 500 grams (23 vs. 20 percent)! French realities make it clear that gradual shifts, driven by price and health concerns, would make it quite realistic to aim at long-term average per capita intakes of no more than 250–300 grams of red meat per week—that is the equivalent of 30–35 kilograms of carcass weight per year, slightly below the recent Japanese level, and only about half of the EU level of the early 2020s.[41]

Changing average intakes of animal protein to lessen the environmental impacts of meat production should involve more than the continuation of the shift from red meat (and especially beef) to chicken. Eggs are now produced with feed conversion efficiencies comparable to chicken, and aquacultured fish (especially herbivorous species) produce the least feed-energy-intensive animal protein.

How can we eat beef better?

How much beef could we produce worldwide without any grain feeding (and the associated environmental costs) and, to reduce land use, with grazing taking place on only a fraction of currently stocked grasslands? There is no agreement about the actual extent and degree of the degradation of grazing lands by excessive use, with estimates ranging from 20 percent of the world's pastures to more than 70 percent of rangelands in arid climates.[42] Consequently, it would be a radical intervention to take half of all pastures completely out of production to allow for their gradual regeneration, and then (to prevent additional deterioration) to lower the stocking on the remaining rangeland (roughly 1.75 billion hectares) to only about half a livestock unit per hectare. This unit is equivalent to 250 kilograms of cattle (live weight), and that rate would be no higher than the typical stocking in sub-Saharan Africa, only about half of the Brazilian rate, and just a quarter of the maximum mandated in the EU.[43]

These assumptions (with a 10 percent culling rate and a 60 percent conversion rate from live weight to carcass weight) would result in an annual production of 10–15 million tons of grass-fed beef. To this, we must add beef that could be grown by the combination of residual feeds—the biomass resulting from harvesting and processing annual crops—and leguminous forages grown in rotations with staple crops. Calculating this total of available feed requires a chain of assumptions, detailed below so any critics can do their own calculations.*

The combination of restricted, low-density grazing and feeding with residual and crop-processing biomass and forage crops could yield anywhere between 20 and 30 million tons of beef (carcass weight) a year, or roughly a third to a half of today's annual total. To reiterate: even when not exploiting more than half of the world's pastures, without using any feed grains (cereal and leguminous), and

---

* They start with typical straw/grain ratios (now usually 1 for rice and corn and 1.3 for wheat) and the shares of crop and milling and oil-pressing residues available for feeding.[44] Milling residues are about 30 percent of the harvested mass for rice and 15 percent for wheat. These two staples produce annually nearly 300 million tons of feed, and pressing oil seeds leaves behind a similar mass of protein-rich oil cakes (20–25 percent of harvested seeds). These are excellent feeds, as is distillers' grain, whey from dairies, and residues from fruit and vegetable canning, and if they were not turned to meat they would have to be disposed of. In contrast, most crop residues (straws and stalks) are not fed to animals; they are either returned to soils to recycle nutrients, protect against erosion, enhance soil's organic matter, and retain moisture, or they are burned. Depending on the assumed rates of residual feeding and on the choice of average feed-to-live-weight conversion ratios, we can end up with a rather wide range of outcomes, but even highly conservative assumptions would end up at least matching the total of grass-fed beef.

relying only on residual and limited forage crop phytomass, we still could produce every year at least on the order of 3 kilograms of beef for every person—compared to the recent mean of about 7.5 kilograms—while reducing the meat's environmental impact by 60–70 percent. But such a reduction of average supply would mean that some people would eat significantly less red meat than they are used to, and there would be financial losses for producers: how far and how fast such shifts could take place remains uncertain.

Maintaining the recent level of pork production (about 120 million tons a year), doubling egg output (which should not be difficult, as we nearly tripled it between 1990 and 2020), and doubling the output of chicken meat (as we did in just two decades between 2000 and 2020) would mean that by 2050 we would have about 150 million tons of red meat, 240 million tons of chicken meat (all carcass weight), and 175 million tons of eggs. And even if the recent wild fish capture remained unchanged (to prevent overfishing and allow for stock recovery) and if aquacultural production rose by 50 percent (it is heading in that direction), we would harvest annually about 220 million tons of fish and crustaceans. This combined harvest would add up to no less than 22 grams of animal protein per day for every one of 9 billion people, matching the recent rate of animal protein supply from meat, eggs, and marine species.

This is just one of many scenarios envisaging adequate animal protein supply produced with substantially reduced environmental impacts: this is not a forecast but an indicator of realistic possibilities that could combine less-than-drastic reductions of red meat consumption with a considerable reduction of environmental impacts. This kind of exploratory calculation is useful for getting an approximate feel for the magnitude of the challenge and for delimiting realistic possibilities, but (despite decades of intensifying globalization) substantial differences and major inequalities remain the norm—for example, food production in China vs. sub-Saharan Africa.

Inequalities

The choice of these two regions should be obvious: during the past three decades no country has demonstrated better what can be done to improve a nation's food supply than China, and during the same period no region has rivaled sub-Saharan Africa as the chronic performance laggard—even as it continues to add to its already large population.

I must stress that for decades I have tried to avoid any unequivocal (quantitative or qualitative) forecasts, and I will not start offering them now. Instead, I always try to offer measured assessments of realities and suggest what might be the most likely outcomes. With long-term food production prospects, I faced this task nearly three decades ago when in 1995 Lester Brown, an American environmentalist and founder of the Worldwatch and Earth Policy institutes, published *Who Will Feed China? Wake-Up Call for a Small Planet*.[45] His slim book argued that China's grain output had already peaked, that it would decline by at least 20 percent by 2030, that the country would soon lose the capacity to feed itself, and that it would face a huge (larger than 300 million tons a year) deficit: simply, that China "could starve the world."

In a reply to this dubious analysis, I noted that many worrying trends in China's recent agricultural development (farmland losses, declining soil, problems with irrigation water) must be considered, along with the fact that the country's cropland area was substantially larger than indicated by official records (the correction was officially made only in the year 2000), that there were some substantial yield gaps to be reduced, and that there were many opportunities for improving nitrogen and water use efficiencies as well as modernizing animal feeding. My conclusions were unequivocal:

> There do not seem to be any insurmountable biophysical reasons why China should not continue feeding itself during the first quarter of the next century. In addition, meeting this challenge does not

require reliance on as yet unproven bioengineering advances or on unprecedented social adjustments. A combination of well-known and well-proven economic and technical fixes—better management, better pricing, better inputs and better environmental protection—can extract enough additional food . . .[46]

And, indeed, between 1995 and 2022, as China's population grew by more than 16 percent, the country's grain harvest rose by nearly 50 percent, and its average per capita food supply has remained well above 3,000 kilocalories a day, closely approaching the rates for the world's most affluent nations.[47]

As noted in the introduction, nearly three decades later I faced an even more extreme forecast: in May 2022 George Monbiot, British writer and political activist, claimed that "the global food system is beginning to look like the global financial system in the run-up to 2008. While financial collapse would have been devastating to human welfare, food system collapse doesn't bear thinking about."[48] But was the global food system really in any imminent pre-collapse state? Thinking so was based on a preposterous analogy. To begin with, many more poorly managed banks folded between 2009 and 2011 than in 2008, and they did so without any crippling consequences for subsequent economic growth, either in the countries of their domicile or on the global level. There has been no collapse of any national or global banking or any exceptional retreat of the global economy: its annual expansion continued at more than 3 percent a year during the second decade of the 21st century.[49]

More fundamentally, any suggestion that the entire global food harvest, spread over a third of the planet's surface and producing annually billions of tons of new plant and animal biomass in environments ranging from the tropics to latitudes above 50°N, could be subject to a near-instant planet-wide collapse is utterly indefensible—unless, of course, we were to experience a truly global catastrophe, be it nuclear war or the Earth's collision with an asteroid. Such claims may make headlines, even sell books, but they are not in the realm of serious analysis. Moreover, if the entire global food system were about to collapse, then I cannot envisage any rapid remedial action able to prevent this.

A serious take on China

In contrast, I can make some informed comments about China's food outlook. These prospects depend on what kind of assumptions are made about the country's future food demand. As already noted, its per capita food supply has been recently well above 3,000 kilocalories a day, and the UN's medium variant forecast expects its population to decline slightly from 1.439 billion in 2020 to 1.402 billion by 2050. The minimalist scenario of China's future food production would see just marginal output increases aimed mainly at improving food intake in the poorest rural regions. A variety of moderate-demand growth projections (mainly due to higher consumption of meat and fruit) would have China's demand for staple grain peaking within 10–15 years to above the recent level, while assumptions of continued significant dietary transition (more meat, more seafood, and more fruit) have the country's food demand up by as much as a third by 2050.[50]

China's future meat consumption will make the greatest difference. The country is by far the world's largest meat consumer (27 percent of the total, twice the US rate), but in per capita terms it is at half the US level. Public opinion polls show that about the same shares of Americans and Chinese (60 percent and 57 percent) consider themselves traditional consumers (regular meat-eaters), while the remainder describe themselves as conscious consumers, ranging from vegans to people deliberately restricting their meat intake.[51] The proportion of conscious meat-eaters is likely to grow, and when combined with the aging population (more health-conscious and tending to eat less meat) it might bring stabilization, and later perhaps even a slight decline of average meat consumption.

The country's environmental conditions must also be considered. China has essentially no room for expanding its cultivated area and cannot afford the continuation of the rapid and large post-1990 conversions of prime suburban farmland to industries and urbanization. Water supply in northern provinces has always been precarious, and the national water balance could get significantly more erratic due to climate change; and (as noted in the fifth chapter) about a fifth of all cropland has been affected by the accumulation of heavy metals. These realities, alongside continued dietary shifts, could lead to substantially higher grain imports, with the overall food self-sufficiency ratio expected to fall from still a very high share of 95 percent in 2016.

A detailed analysis of rice production potential in China (considering specific provincial performances) concluded that the country is likely to remain self-sufficient in this staple even if it can do nothing more than sustain current yield and consumption trajectories and if it will not reduce the existing production area. That study also found that the greatest return on investment would come from increased yields of double-rice systems in general, and closing relatively large yield gaps in three single-rice provinces.[52]

The chances are that the conclusions I made nearly three decades ago can be extended, with some caveats, for another two to three decades. Unless the country becomes as carnivorous (per capita) as the US, it should not put any excessive claims on the global grain market and it should be able to manage (mostly, if not completely) by slightly increasing its recent levels of food production and food imports. Further dietary changes may lead to higher imports of feed grains, cooking oils, and specialty products. These developments might be a major contributor to higher prices, whose impact would be felt by low-income countries that, unlike China with its large foreign trade surpluses, would be much less able to pay for them. So, no food apocalypse in or caused by China anytime soon.

Food production in sub-Saharan Africa

The prospects are much more worrisome for sub-Saharan Africa. The *Global Report on Food Crises* for 2022 listed 35 distressed nations:

four in Central America, six in Asia, but 25 in sub-Saharan Africa, which is (leaving aside small island countries) nearly 60 percent of the region's territories.[53] As I noted in the sixth chapter, no other region has such low staple crop yields, and no other region has such a need to narrow the yield gaps, because a quarter of its population is experiencing permanent food insecurity (this increases in dry seasons, or periods of drought or locust infestation) and because it will become the home of about half of the world's additional population by 2050. All sub-Saharan crops yield well below their potential, with yield gaps present not only for cereal staples (corn, rice, sorghum) but also for potatoes and for regionally important legumes. And in all cases (as the following key examples will illustrate), it is a shortage of macronutrients, above all nitrogen, that presents the greatest obstacle.

A recent study estimated that the region's corn yields have to increase from the existing level, equal to about 20 percent of their water-limited yield potential, to 50–75 percent of their potential, but that minimum nitrogen fertilizer inputs must rise 9- to 15-fold, otherwise continuous soil nutrient "mining" will lead to even lower yields.[54] Smallholders (with less than 2 hectares of cropland) apply often less than 10 kilograms of nitrogen per hectare (plus variable, but always insufficient, amounts of organic fertilizer), leading to a gradual decline of soil nitrogen content and to chronically low yields. Trials with different rice production systems (irrigated, rainfed, upland rice) in 17 countries showed nitrogen to be the most limited nutrient, followed by phosphorus, and the yields that rose with nitrogen applications were still well below the Asian maxima.[55] Another study in Eastern and Southern Africa found that yield gaps of roughly 1–3 tons per hectare were also related to weed suppression, bird control, land-leveling, and straw management (recycling, burning in field, removing), emphasizing that higher fertilization alone will not suffice overall crop management.[56]

But one major factor complicates all future corn yield scenarios: uncertainty about the region's prevailing rooting depths. Unlike the relatively young and deep soils of the US Corn Belt, agricultural soils in sub-Saharan Africa have restricted rooting zones (due to the presence of laterite, an aluminum- and iron-rich soil layer) as well as limited water retention capacity. That is why an assessment based on the best available spatial information concluded that sub-Saharan Africa could produce modest corn surpluses only if its rootable soil depths were comparable to those of the US Corn Belt or Argentinian Pampas.[57] Because of the high sensitivity to this single variable, more realistic assessments of potential yield gains need better information of soil properties for most of the region.

Another constraint is a high year-to-year variability in yields due to normal variation of rainfall: in sub-Saharan Africa it is more than double that in the Corn Belt, and the ensuing higher uncertainty in yield expectations makes it, obviously, more risky to invest in higher fertilizer applications because they may not bring profitable returns during very dry years.

Corn is not the only staple underperformer. Potato yields in sub-Saharan Africa now average just above 10 tons per hectare, while the potential yield is above 60 t/ha and yields in research plots are in excess of 30 t/ha. The main reasons for a 50-ton yield gap are poor seed quality, the presence of bacterial wilt, poor soil quality, inadequate nutrient supply, and insect pests.[58] Cowpea is a leading leguminous grain in West Africa but its yields remain low, mostly between 500 and 800 kilograms per hectare.[59] Yet again, increasing the supply of the three macronutrients remains the key factor, with the control of insect pests coming second. As with all legumes, cowpea can fix its own nitrogen, but this symbiotic activity is limited by phosphorus availability, and the crop responds to urea applications.

The need for higher fertilizer applications is an omnipresent factor, and its importance was confirmed by an analysis of household data in eight of the region's countries: yield differences are reduced with higher fertilizer use, and as expected this is particularly the case when these applications are combined with improved seeds.[60] Also as expected, yield gap decreases significantly when farmers have access to information on optimal production practices, especially in regions with low yield potential. Other notable findings were that yield gap discriminates against female-headed households, that poverty gaps increase with yield gaps, and the yield gap for smallholders increases with the cultivated area and begins to decline only in farms larger than 3.3 hectares.

What the future holds

The dominant reasons for yield gaps are clear and remedies are well known, but the critical question remains: can the region feed itself by 2050 if the needed yield-gap-closing steps are enacted? An international group of agronomists (with contributors from Africa, Europe, and the US) published a yield gap analysis of 10 sub-Saharan countries and concluded that "it will not be feasible to meet future sub-Saharan Africa cereal demand on existing production area by yield gap closure alone." [61] This verdict rested on realistic examinations of yield gap in the region's 10 major countries (based on location-specific data), and it led the authors to suggest that additional measures—above all, increased cropping intensity (double- or triple-cropping) and the expansion of irrigated cropland—will be needed in order to prevent further cropland expansion (deforestation and grassland conversions) and avoid increasing the region's dependence on staple grain imports.

This does not mean that by the middle of the 21st century sub-Saharan Africa could not achieve greater food security while accommodating another billion people, but that doing so will require complex and sustained efforts that must go far beyond the reliance on improving harvests from existing fields, and that substantial food imports from the Americas and Eurasia will still be needed. Few of today's global challenges are more important than making sure that Africa will, at least, narrow the production–demand gap. Fundamental reasons (beyond the outlined natural constraints) for lagging behind have been obvious for decades: poor governance, widespread political instability, seemingly endless cross-border conflicts, and civil wars in too many countries (Sudan, South Sudan, Eritrea, Ethiopia, Somalia, Rwanda, Burundi, DRC, Angola, Nigeria, Niger, Mali, Liberia, Sierra Leone, Mozambique), and intra-national tensions (chronic in the continent's two most populous nations, Ethiopia and Nigeria), combined with deep economic inequalities and excessive reliance on imports. And failures to remove, or at least to lower, these basic obstacles to economic progress could only worsen the worrisome food-supply situation that, in a circular fashion, could deepen some of these chronic crises.

Global warning

Then there is, of course, the even greater challenge arising from the interaction of global warming with food production. This involves many undesirable developments ranging from changes in the global water cycle (there will be more rain and snow in a warmer world, but not necessarily in places that need it most) to shifts in growing seasons (earlier dates of French wine grape harvests are an excellent documentation of this shift) to effects on the nutritional content of staple crops.[62] But the rising atmospheric concentrations of CO_2 have also benefitted plants, especially crops, and have resulted in indisputable biospheric greening.

Satellite monitoring shows that during the first two decades of the 21st century up to half of the world's vegetated land showed significant greening (denser foliage observed from space) and that this increase in the leaf area of plants (crops, grasslands, and trees) was equivalent to the area covered by all the Amazon rainforests; in contrast, less than 4 percent of the Earth showed browning—and a study of recent corn yields in favorable environments found that nearly half of that gain was associated with a decadal climate trend, compared to about 40 percent with agronomic improvements and just 13 percent with improvements in genetic yield potential.[63] Moreover, the greening effect may be even more pronounced in the future, because the existing models of CO_2 absorption by plants do not reflect the latest physiological understanding—including plant acclimation to rising temperatures, changes due to the speed with which CO_2 moves to the sites containing Rubisco, and the redistribution of leaf nitrogen. Inclusion of these effects indicates that the global greening may be up to 20 percent greater during the closing decades of the 21st century than the past models suggest.[64]

Our understanding of these complexities and interactions continues to evolve and to improve, but the most detailed examinations of these effects rely on long-range forecasting models that (as just indicated by recent revisions of the extent of the future greening) are subject to the usual problems, ranging from basic assumptions and limits imposed by our understanding of the complexities of climate change and its interaction with photosynthetic productivity, to the unknown extent of mitigation and adaptation steps we will take during the coming decades in order to manage the expected impacts. These range from modifying planting dates, shortening crop maturation periods, and expanding double- and where possible even triple-cropping, to the introduction of more drought-tolerant cultivars and adopting significant dietary adjustments.

And, inevitably, rising temperatures and increasing CO_2 concentrations will have substantial plant-specific and regional differences. One recent model predicts that by 2050 a high-emissions scenario could cut corn yields in China and Brazil by as much as 1.5–2 t/ha, and rice yields in Asia and Latin America by 0.5–1 t/ha— but increase the wheat yields in higher latitudes by 1–2 t/ha.[65] And if such a yield drop (at this time modeled, not actual) seems worrisome, then another recent global assessment concluded that by optimizing crop cultivation (changing to crops with the highest regional yield potential) could produce enough food for an additional 825 million people—while reducing the total water need by 10 percent.[66] And yet another study showed that the timely adaptation of growing periods to climate change (adjusting sowing and harvesting dates) would reduce the negative impacts of climate change and increase actual global crop yields by about 12 percent.[67]

And even if these welcome gains would not be as pronounced as they have been modeled, we should (as already argued) adjust our final uses of cereal grains: we do not need to feed more than a third of them to animals and we do not have to convert nearly 10 percent of them into biofuels. And there is yet another important factor that will affect global food production on the demand side: population aging. While both the basal metabolism and the total energy expenditure remain fairly stable between the ages of 20 and 60 years, they begin to decline after that, with reduced energy expenditure (lower activity) being more important than the loss of body mass.[68] Not surprisingly, in Japan, where in 2022, 35 percent of people were older than 60, average per capita food supply has already declined by about 10 percent since 1990, and similar gradual reductions can be expected in other rapidly aging nations, a group that now includes China.

As a result, I remain agnostic about long-term prospects of the global food supply: they are not foreordained because of the continuing population growth or because of continued global warming, and they will not be limited by what we can do today.

But that would be a topic for another, and quite a lengthy, book. This one was predominantly about the basic biophysical determinants (photosynthetic efficiency, nutrient needs, feeding conversion rates) and the outcomes and uses of food production (nutrition requirements, food supply, and the nature of the global food system). That was topped by brief assessments of promised transformations that (in my judgment) do not have a great chance of fundamentally changing the global food supply by 2050, and measures that would work and whose assiduous pursuit might result in us coming closer to that ideal objective of producing enough food for humanity and doing so with reduced environmental consequences.

My goal has been to provide a deeper understanding of the fundamental limitations and inherent complexities of global food production. Foremost among these are the biophysical limits on photosynthetic productivity and the recalcitrance of changing human behavior, especially where prevailing diets are concerned. These realities reduce the scope of options in coming up with solutions to global food security that could be assured without further large-scale environmental degradation. But despite this narrowing of choices, a realistic outlook for maintaining adequate global food supply remains promising. No unprecedented gains and no untried radical solutions are required to provide the next generation with adequate food supply: we just need to keep on improving production efficiencies, reducing waste, adjusting diets, and promoting measures that reduce food's overall environmental impact.

Declining populations (now the norm in all high-income countries that do not allow large-scale immigration), falling fertility rates (which includes all populous countries, with Asian rates already below or near the replacement level, but with African countries still well above it), and moderating demand for food (typical of all aging societies) should make these quests easier than we thought a generation ago, when the best long-range forecasts saw the global population rising well above 10 billion people. Now the most likely outlook is for a peak of 9.7 billion by the mid-2060s, followed by a decline to about 8.8 billion in 2100, a total 2 billion lower than some previous forecasts.[69]

At the same time, the unequally distributed climatic consequences of global warming will bring a combination of negative and positive changes, and hence the quest will become harder in some regions and for some crops. And while the potential for radical gains due to the genetic modification of crops and animals looks impressive, the dates of large-scale commercial adoption remain unknown, as do the prospects for any substantial voluntary modifications of prevailing diets. Global transitions take a long time.

There is nothing new in facing such combinations of uncertainties. I wrote this book to offer facts rather than speculations, and as it happens, the facts are reassuring. We should remain suspicious of any long-range quantitative forecasts, but—considering basic biophysical realities, continuing performance gains, and a realistic assessment of future improvements—it is rational to argue that, barring mass-scale conflicts and unprecedented social breakdown, the world will be able to feed its growing population beyond the middle of the 21st century, when the combination of new demographic realities and new scientific advances may present us with entirely new options.

References and Notes

1. What Did Agriculture Ever Do for Us?

1 Chimpanzee scavenging has been well documented, but it is rare, as are the instances of cannibalism: Watts, D.P. 2008. "Scavenging by chimpanzees at Ngogo and the relevance of chimpanzee scavenging to early hominin behavioral ecology." *Journal of Human Evolution* 54: 125–33; Goodall, J. 1977. "Infant killing and cannibalism in free-living chimpanzees." *Folia Primatologica* 28: 259–89; Nishie, H. and M. Nakamura. 2017. "A newborn infant chimpanzee snatched and cannibalized immediately after birth: Implications for 'maternity leave' in wild chimpanzees." *American Journal of Biological Anthropology* 165(1): 104–9. For the best evidence of Neanderthal cannibalism see: Rougier, H. et al. 2016. "Neandertal cannibalism and Neandertal bones used as tools in Northern Europe." *Scientific Reports* 6: 29005.

2 Here are just a few notable examples of chimpanzee diet studies: Wrangham, R.W. and E.Z.B. Riss. 1990. "Rates of predation on mammals by Gombe chimpanzees, 1972–1975." *Primates* 31: 157–70; Basabose, A.K. 2002. "Diet composition of chimpanzees inhabiting the montane forest of Kahuzi, Democratic Republic of Congo." *American Journal of Primatology* 58: 1–21; Watts, D.P. et al. 2012. "Diet of chimpanzees (*Pan troglodytes schweinfurthii*) at Ngogo, Kibale National Park, Uganda, 1. Diet composition and diversity." *American Journal of Primatology* 74: 114–29; Piel, A.K. et al. 2017. "The diet of open-habitat chimpanzees (*Pan troglodytes schweinfurthii*) in the Issa valley, western Tanzania." *Journal of Human Evolution* 112: 57–69; Moore, J. et al. 2017. "Chimpanzee vertebrate consumption: Savanna and forest chimpanzees compared." *Journal of Human Evolution* 112: 30–40.

3 Pruetz, J.D. and P. Bertolani. 2007. "Savanna chimpanzees, *Pan troglodytes verus*, hunt with tools." *Current Biology* 17: 412–17.

4. Wessling, E.G. et al. 2020. "Chimpanzee (*Pan troglodytes verus*) density and environmental gradients at their biogeographical range edge." *Journal of Primatology* 41:822–48; Chitayat, A.B. et al. 2021. "Ecological correlates of chimpanzee (*Pan troglodytes schweinfurthii*) density in Mahale Mountains National Park, Tanzania." *PLoS ONE* 16(2): e0246628.
5. Pobiner, B.L. 2020. "The zooarchaeology and paleoecology of early hominin scavenging." *Evolutionary Anthropology* 29: 68–82.
6. Ben-Dor, M. et al. 2021. "The evolution of the human trophic level during the Pleistocene." *Yearbook of Physical Anthropology* 175 (suppl. 72): 27–56.
7. For the history of the overkill hypothesis see: Martin, P.S. 1958. "Pleistocene ecology and biogeography of North America." *Zoogeography* 151: 375–420; Martin, P.S. 2005. *Twilight of the Mammoths*. Berkeley: University of California Press. For its critique see: Smil, V. *Harvesting the Biosphere*. Cambridge, MA: MIT Press, 78–87.
8. Ethnographic studies of foragers are reviewed and summarized in: Murdock, G.P. 1967. "Ethnographic atlas." *Ethnology* 6: 109–236; Kelly, R.L. 2013. *The Lifeways of Hunter-Gatherers: The Foraging Spectrum*. Cambridge: Cambridge University Press; Cummings, V. et al., eds. 2018. *The Oxford Handbook of the Archaeology and Anthropology of Hunter-Gatherers*. Oxford: Oxford University Press.
9. Marlowe, F.W. 2005. "Hunter-gatherers and human evolution." *Evolutionary Anthropology* 14: 54–67.
10. Maschner, E.D.G. and B.M. Fagan. 1991. "Hunter-gatherer complexity on the west coast of North America." *Antiquity* 65: 921–3; Ames, K.M. 1994. "Complex hunter-gatherers, ecology, and social evolution." *Annual Review of Anthropology* 23: 209–29.
11. Huffa, C.D. et al. 2010. "Mobile elements reveal small population size in the ancient ancestors of Homo sapiens." *Proceedings of the National Academy of Sciences* 107: 2147–52.
12. Tallavaaraa, M. et al. 2015. "Human population dynamics in Europe over the Last Glacial Maximum." *Proceedings of the National Academy of Sciences* 112: 8232–7.

13 Bailey, R.C., G. Head, M. Jenike et al. 1989. "Hunting and gathering in tropical rain forest: Is it possible?" *American Anthropologist* 91: 59–82; Bailey, R.C. and T.N. Headland. 1991. "The tropical rain forest: Is it a productive environment for human foragers?" *Human Ecology* 19: 261–85.

14 Sheehan, G.W. 1985. "Whaling as an organizing focus in Northwestern Eskimo society." In T.D. Price and J.A. Brown, eds., *Prehistoric Hunter-Gatherers*. Orlando, FL: Academic Press, 123–54; Krupnik, I.I. and S. Kan. 1993. "Prehistoric Eskimo whaling in the Arctic: Slaughter of calves or fortuitous Ecology?" *Arctic Anthropology* 3: 112.

15 Childe, V.G. 1936. *Man Makes Himself*. London: Watts & Company, 61. On long coexistence of agriculture and foraging see: Smil, V. 2017. *Energy and Civilization: A History*, Cambridge, MA: MIT Press; Bharucha, Z. and J. Pretty. 2010. "The roles and values of wild foods in agricultural systems." *Philosophical Transactions of the Royal Society B* 365: 2913–26.

16 Lo Cascio, E. 1994. "The size of the Roman population: Beloch and the meaning of the Augustan census figures." *Journal of Roman Studies* 84: 23–40; Scheidel, W. 2007. *Roman Population Size: The Logic of the Debate*. Stanford, CA: Princeton/Stanford Working Papers in Classics.

17 Zeder, M. 2006. "Central questions in the domestication of plants and animals." *Evolutionary Anthropology* 15: 105–117. Hodder explains this reverse explanation as the process of human–thing entanglement: Hodder, I. 2012. *Entangled: An Archaeology of the Relationships between Humans and Things*. Hoboken, NJ: John Wiley.

18 Richerson, P.J. et al. 2001. "Was agriculture impossible during the Pleistocene but mandatory during the Holocene? A climate change hypothesis." *American Antiquity* 66: 387–412.

19 Binford, L.R. 2001. *Constructing Frames of Reference: An Analytical Method for Archaeological Theory Building Using Ethnographic and Environmental Data Sets*. Berkeley, CA: University of California Press.

20 Butzer, K.W. 1976. *Early Hydraulic Civilization in Egypt*. Chicago: University of Chicago Press; Butzer, K.W. 1984. "Long-term Nile flood variation and political discontinuities in Pharaonic Egypt." In: J.D.

Clark and S.A. Brandt, eds., *From Hunters to Farmers*. Berkeley, CA: University of California Press, 102–12.

21. Buck, J.L. 1937. *Land Utilization in China*. Nanking: University of Nanking; Perkins, D.S. 1969. *Agricultural Development in China, 1368–1968*. Chicago: University of Chicago Press.

22. Chorley, G.P.H. 1981. "The agricultural revolution in Northern Europe, 1750–1880: Nitrogen, legumes, and crop productivity." *Economic History Review* 34: 71–93; Clark, G. 1991. "Yields per acre in English agriculture, 1250–1850: Evidence from labour inputs." *Economic History Review* 44: 445–60; Bieleman, J. 2010. *Five Centuries of Farming: A Short History of Dutch Agriculture*. Wageningen: Wageningen University.

23. Here are the 2021 rounded global totals (in billions of hectares) according to the FAO. https://www.fao.org/faostat/en/#data/RL: agricultural land 4.82; cropland 1.58; arable land 1.40; perennial crops 0.18; pastures and meadows 3.20. The quotient is calculated by dividing the global population of 7.8 billion by total cropland.

24. FAO. 2021. *The State of Food Security and Nutrition in the World*. https://www.fao.org/faostat/en/#data/RL

25. FAO. 2001. *Human Energy Requirements*. Rome: FAO; WHO. 2007. *Protein and Amino Acid Requirements in Human Nutrition: Report of a Joint FAO/WHO/UNU Expert Consultation*.

26. Perhaps the best online source of detailed nutritional values (energy, macronutrients, micronutrients) is https://www.nutritionvalue.org. All data can be accessed per unit of weight (international and US measures) or volume (US cups) as well as per serving (a single, size-dependent, piece where applicable); moreover, it has data for raw as well as processed and prepared foods.

27. During the early 2020s more than 70 percent of all figs were produced in just five countries: Turkey, Morocco, Egypt, Algeria, and Iran.

28. And also more than the combined annual harvest of the world's five most important fruit species: bananas, watermelons, apples, oranges, and grapes.

29. Rothman, J. et al. 2007. "Nutritional composition of the diet of the gorilla (*Gorilla beringei*): a comparison between two montane habitats."

Journal of Tropical Ecology 23: 673–82; Schulz, D. et al. 2018. "Anaerobic fungi in gorilla (*Gorilla gorilla gorilla*) feces: an adaptation to a high-fiber diet?" *International Journal of Primatology* 39: 567–80.

30 Furness, J.B. et al. 2015. "Comparative Gut Physiology Symposium: Comparative physiology of digestion." *Journal of Animal Science* 93.

31 Fry, E. et al. 2020. "Functional architecture of deleterious genetic variants in the genome of a Wrangel Island mammoth." *Genome Biology and Evolution* 12: 48–58.

32 Stefansson, V. 1946. *Not by Bread Alone*. New York: Macmillan.

33 Nunavut Department of Health. 2013. *Nutrition Fact Sheet Series Inuit Traditional Foods*. Iqaluit: Nunavut Department of Health.

34 Tucker, A. 2009. "In search of the mysterious narwhal." *Smithsonian Magazine*. https://www.smithsonianmag.com/science-nature/in-search-of-the-mysterious-narwhal-124904726/

35 Shaw, J.H. 1995. "How Many Bison Originally Populated Western Rangelands?" *Rangelands* 17: 148–50; Isenberg, A.C. 2000. *The Destruction of the Bison: An Environmental History, 1750–1920*. Cambridge: Cambridge University Press.

36 Forest Service. 2018. "Hunting, fishing and conservation go hand in hand." https://www.fs.usda.gov/inside-fs/delivering-mission/sustain/hunting-fishing-and-conservation-go-hand-hand

37 Rowell, R.M. et al. 2012. *Handbook of Wood Chemistry and Wood Composites*. Boca Raton, FL: CRC Press; Maleki, S.S. et al. 2016. "Characterization of cellulose synthesis in plant cells." *The Scientific World*.

38 Average global body mass is lowered by still rather high shares of children, and young adults in many low-income countries.

39 Average global cattle body mass is lowered by smaller bodies of animals in India and Brazil, two countries with the largest herds.

40 Dijkstra, J. et al., eds. 2005. *Quantitative Aspects of Ruminant Digestion and Metabolism*. Wallingford: Centre for Agriculture and Bioscience International.

41 Khan, M.A. and A. Wasim, eds. 2018. *Termites and Sustainable Management Volume 1: Biology, Social Behaviour and Economic Importance*. Cham: Springer.

42 Zimmerman, P.R. et al. 1982. "Termites: A potentially large source of atmospheric methane, carbon dioxide, and molecular hydrogen." *Science* 218: 563–5.

43 Yield of 40 t/ha, extracted sugar (after losses) 10 percent of the cut cane; 4 tons of sucrose (at 17 MJ/kg) contain 68 GJ; daily per capita energy requirement of 9.2 MJ adds up to 3.35 GJ/year; 68 GJ/3.35 GJ = 20.2.

44 Walvin, J. 2018. *Sugar: The World Corrupted: From Slavery to Obesity*. New York: Pegasus Books.

45 Of which more than half is produced by just five countries: Brazil, India, China, Thailand, and the United States.

2. Why Do We Eat Lots of Some Plants and Not Others?

1 Barigozzi, C., ed. 1986. *The Origin and Domestication of Cultivated Plants*. Amsterdam: Elsevier; Zohary, D. et al. 2012. *Domestication of Plants in the Old World: The Origin and Spread of Domesticated Plants in Southwest Asia, Europe and the Mediterranean Basin*. Oxford: Oxford Scholarship Online.

2 Armelagos, G.J. and K.N. Harper. 2005. "Genomics at the origins of agriculture, part one." *Evolutionary Anthropology* 14: 68–77; Kantar, M.B. et al. 2017. "The genetics and genomics of plant domestication." *BioScience* 67: 971–82.

3 Damama, A.B. et al., eds. 1998. *The Origins of Agriculture and Crop Domestication*. Berkeley: University of California.

4 Hillman, G.C. and M.S. Davies. 1990. "Domestication rates in wild-type wheats and barley under primitive cultivation." *Biological Journal of the Linnean Society* 39: 39–78.

5 Meyer, R.S. et al. 2012. "Patterns and processes in crop domestication: an historical review and quantitative analysis of 203 global food crops." *New Phytologist* 196: 29–48.

6 Biribá's English name says it all: the lemon meringue pie fruit of South America is similar to a more common custard apple. Every part of a small huauzontle plant (its branches, leaves, flowers, and seeds) can be

eaten: this Mexican vegetable and herb is especially popular in the country's central region. Noni, a light green, bitter, and smelly fruit, was usually eaten in Southeast Asia only when other food supplies fell short.

7 FAO. 2022. "FAOSTAT: Crops and Livestock Products." https://www.fao.org/faostat/en/#data/QCL

8 Production of US corn ethanol now claims about 40 percent of the crop's annual harvest, while about 45 percent of Brazil's sugarcane harvest has been recently converted to ethanol: US Department of Agriculture. 2021. "Feedgrains Sector at a Glance." https://www.ers.usda.gov/topics/crops/corn-and-other-feed-grains/feed-grains-sector-at-a-glance/; Barros, S. 2021. "Sugar Semi-annual."

9 Brazil's sugar mills process annually more than 600 million tons of cane, and that yields about 160 million tons of bagasse: Tilasto. 2021. https://www.tilasto.com/en/country/brazil/energy-and-environment/bagasse-production

10 Weiss, E. and D. Zohary. 2011. "The Neolithic Southwest Asian founder crops: Their biology and archaeobotany." *Current Anthropology* 52: S237–S254.

11 Zaharieva, M. et al. 2010. "Cultivated emmer wheat (*Triticum dicoccon* Schrank), an old crop with a promising future: a review." *Genetic Resources and Crop Evolution*.

12 Einkorn.com. 2022. "The history of einkorn, nature's first and oldest wheat." https://www.einkorn.com/einkorn-history/

13 Caracuta, V. et al. 2015. "The onset of faba bean farming in the Southern Levant." *Scientific Reports* 5: 14370.

14 Sedivy, E.J. et al. 2017. "Soybean domestication: the origin, genetic architecture and molecular bases." *New Phytologist* 214: 539–53.

15 Swarts, K. et al. 2017. "Genomic estimation of complex traits reveals ancient maize adaptation to temperate North America." *Science* 357: 512–5.

16 Haas, M. et al. 2019. "Domestication and crop evolution of wheat and barley: Genes, genomics, and future directions." *Journal of Integrative Plant Biology* 61: 204–25.

17 By far the best source for everything concerning bread is: Myhrvold, N. and F. Migoya. 2017. *Modernist Bread*. Bellevue, WA: The Cooking Lab.

18 D'Andrea, S. K. et al. 2007. "Early domesticated cowpea (*Vigna unguiculata*) from Central Ghana." *Antiquity* 81: 686–98.
19 Henricks, R.G. 1998. "Fire and rain: A look at Shen Nung 神農 (The Divine Farmer) and his ties with Yen Ti 炎帝 (The Flaming Emperor or Flaming God)." *Bulletin of the School of Oriental and African Studies* 61: 102–24.
20 For the composition of foods see: https://www.nutritionvalue.org
21 This is more than the recommended requirements based on the prevailing body masses and activity levels of the US population. For their calculation see: FAO. 2001. *Human Energy Requirements*. Rome: FAO.
22 And yet millions of people were forced to survive on such rations. The NKVD (the Soviet National Committee of Internal Security) Order 00943 of August 14, 1939 (*On the introduction of new standards of nutrition and clothing rations for prisoners in the correctional labour camps and colonies of the NKVD of the USSR*) specified that those prisoners who fell behind their production quotas, and the disabled, would get 600 grams of rye bread and 100 grams of kasha (buckwheat porridge), as well as 500 grams of potatoes and vegetables and 30 grams of meat—but the punishment ration was just 400 grams of bread, 35 grams of kasha, 400 grams of potatoes and vegetables, and no meat. "European Memories of the Gulag." 2022. "Food rations." https://www.gulagmemories.eu/en/sound-archives/media/food-rations
23 Joint FAO/WHO/UNU Expert Consultation on Protein and Amino Acid Requirements in Human Nutrition. 2007. *Protein and Amino Acid Requirements in Human Nutrition: Report of a Joint FAO/WHO/UNU Expert Consultation*. Rome: FAO.
24 Report of an FAO Expert Consultation. 2013. *Dietary Protein Quality Evaluation in Human Nutrition*. Rome: FAO; Mathai, J.K. et al. 2017. "Values for digestible indispensable amino acid scores (DIAAS) for some dairy and plant proteins may better describe protein quality than values calculated using the concept for protein digestibility-corrected amino acid scores (PDCAAS)." *British Journal of Nutrition* 117: 490–9.
25 Processing (wet-milling and coagulation to make bean curd, *doufu*) and fermentation (to make *sufu*) made soybeans easier to digest: Han, B.

et al. 2001. "A Chinese fermented soybean food." *International Journal of Food Microbiology* 65: 1–10.

26 Dillehay, T.D. et al. 2007. "Preceramic adoption of peanut, squash, and cotton in northern Peru." *Science* 316: 1890–3.

27 Buck, J.L. 1930. *Chinese Farm Economy*. Nanking: Nanking University Press; Buck, J.L. 1937. *Land Utilization in China*. Nanking: Nanking University Press.

28 Bozhong, L. and P. Li. 1998. *Agricultural Development in Jiangnan, 1620–1850*. New York: St. Martin's Press, 111; Li, L.M. and A. Dray-Novey. 1999. "Guarding Beijing's food security in the Qing dynasty: State, market, and police." *The Journal of Asian Studies* 58: 992–1032.

29 Oddy, D. 1970. "Food in nineteenth century England: Nutrition in the first urban society." *Proceedings of the Nutrition Society* 29: 150–7.

30 National fortification guidelines vary, but in most jurisdictions refined (highly milled) wheat flour is now enriched with thiamine (vitamin B1), riboflavin (vitamin B2), niacin (vitamin B3), folic acid, and, to prevent anemia, iron.

31 FAO. 2022. "FAOSTAT Trade: Crops and Livestock Products." https://www.fao.org/faostat/en/#data/TCL

32 Smil, V. 2004. *China's Past, China's Future*. London, Routledge, 91.

33 Smil, V. and K. Kobayashi. 2012. *Japan's Dietary Transition and Its Impacts*. Cambridge, MA: MIT Press.

34 Bennett, M.K. 1935. "British wheat yield per acre for seven centuries." *Economy and History* 3: 12–29; Stanhill, G. 1976. "Trends and deviations in the yield of the English wheat crop during the last 750 years." *Agro-ecosystems* 3: 1–10.

35 Rosentrater, K., ed. 2022. *Storage of Cereal Grains and Their Products*. Amsterdam: Elsevier.

36 Kumar, D. and P. Kalita. 2017. "Reducing postharvest losses during storage of grain crops to strengthen food security in developing countries." *Foods* 6 (8).

37 Posner, E.S. and A.N. Hibbs. 2004. *Wheat Flour Milling*. Eagan, MN: Cereals and Grains Association.

38 International Rice Research Institute. 2022. "Milling Yields." http://www.knowledgebank.irri.org/step-by-step-production/postharvest/milling/producing-good-quality-milled-rice/milling-yields

39 Pace, C.M. 2012. *Cassava: Farming, Uses, and Economic Impact.* Hauppauge, NY: Nova Science Publishers.

40 Peñarrieta, J.M. et al. 2011. "Chuño and tunta: The traditional Andean sun-dried potatoes." In: Caprara, C., ed., *Potatoes: Production, Consumption and Health Benefits.* Hauppauge, NY: Nova Science Publishers, 1–12.

41 A comprehensive collection of national crop calendars is available at: FAO. 2022. "Crop Calendar." https://cropcalendar.apps.fao.org/#/home

42 Wong, R.B. et al. 1991. *Nourish the People: The State Civilian Granary System in China, 1650–1850.* Ann Arbor, MI: University of Michigan Press.

43 Erdkamp, P. 2009. *The Grain Market in the Roman Empire: A Social, Political and Economic Study.* Cambridge: Cambridge University Press.

44 Watanabe, S. and A. Munakata. 2021. "China hoards over half the world's grain, pushing up global prices." https://asia.nikkei.com/Spotlight/Datawatch/China-hoards-over-half-the-world-s-grain-pushing-up-global-price

45 Agflows. 2022. "A Guide to Bulk Carriers Types for Agricultural Commodities." https://www.agflow.com/commodity-trading-101/a-guide-to-bulk-carriers-types-for-agricultural-commodities/

46 Diamond, J. 1987. "The worst mistake in the history of the human race." *Discovery* May 1987: 64–6.

47 Lee, R.B. and I. DeVore, eds. 1968. *Man the Hunter.* New York: Aldine Publishing.

48 Sahlins, M. 1968. "Notes on the original affluent society." In: *Man the Hunter*, 85–9.

49 Keeley, L.H. 1997. *War Before Civilization.* Oxford: Oxford University Press; Kaplan, D. 2000. "The darker side of the 'Original affluent society.'" *Journal of Anthropological Research* 56: 301–4; Buckner, W. 2017. "Romanticizing the hunter-gatherer." *Quillette* December 16, 2017.

50 Diamond, "The worst mistake in the history of the human race."
51 D'Ormesson, J. 2016. *The Glory of the Empire*. New York: New York Review Books, 19.
52 Scott, J.C. 2017. *Against the Grain: A Deep History of the Earliest States*. New Haven, CT: Yale University Press.
53 Davis, W. 2015. *Wheat Belly: Lose the Wheat, Lose the Weight, and Find Your Path Back to Health*. New York: Collins.
54 For recent annual average per capita food balances see: FAO. 2022. "Food balances (2010–)." https://www.fao.org/faostat/en/#data/FBS
55 Macrotrends. 2023. "World Life Expectancy 1950–2023." https://www.macrotrends.net/countries/WLD/world/life-expectancy
56 Perry, J.M.G. and S.L. Canington. 2019. "Primate Evolution." https://explorations.americananthro.org/wp-content/uploads/2019/10/Chapter-8-Primate-Evolutio-2.0.pdf

3. The Limit of What We Can Grow

1 Barker, A.V. and D.J. Pilbeam, eds. 2015. *Handbook of Plant Nutrition*. London: Routledge.
2 Wang, A. et al. 2022. "CO_2 enrichment in greenhouse production: Towards a sustainable approach." *Frontiers in Plant Science* 13.
3 If you want to see some astonishing results, try to calculate (among many possible examples) the levels of air pollution we would now be enduring if efficiencies of energy converters had remained at the 1950—or, far worse, at the 1900—levels.
4 Smil, V. 2017. *Energy and Civilization: A History*. Cambridge, MA: MIT Press.
5 Smil, V. 2010. *Prime Movers of Globalization: The History and Impact of Diesel Engines and Gas Turbines*. Cambridge, MA: MIT Press; General Electric, 2022. "GE's HA turbine recognized for powering world's most efficient power plants in both 50hz & 60hz segments." https://www.gevernova.com/gas-power/resources/articles/2018/nishi-nagoya-efficiency-record

6 Energy Star. 2021. "ENERGY STAR Most Efficient 2021—Furnaces," https://www.energystar.gov/products/most_efficient/furnaces

7 Bowers, B. 1998. *Lengthening the Day: A History of Lighting Technology.* Oxford: Oxford University Press; Lumega. 2018. "Highest LED energy efficiency." https://www.lumega.eu/laeringsmiljoe. Here are the comparisons in lumens per watt: incandescent bulbs, 10; fluorescent tubes, 80–100; LEDs, 100–200+.

8 Shockley, W. and H. J. Queisser. 1961. "Detailed balance limit of efficiency of p-n junction solar cells." *Journal of Applied Physics* 32: 510–19.

9 EnBW Company. 2021. "Biggest solar park without state funding inaugurated."https://www.enbw.com/company/press/enbw-inaugurates-germany-s-largest-solar-park.html; National Renewable Energy Laboratory. 2022. "Best Research-Cell Efficiency Chart." https://www.nrel.gov/pv/cell-efficiency.html

10 Benedict, F. and E. Cathcart. 1913. *Muscular Work: A Metabolic Study with Special Reference to the Efficiency of the Human Body as a Machine.* Washington, DC: Carnegie Institute; Lindinger, M.I. and S.A. Ward. 2022. "A century of exercise physiology: key concepts in . . ." *European Journal of Applied Physiology* 122: 1–4.

11 Forseth, I.N. 2010. "The Ecology of Photosynthetic Pathways." *Nature Education Knowledge* 3(10): 4.

12 Encyclopedia of the Environment. 2022. RubisCO. https://www.encyclopedie-environnement.org/en/zoom/rubisco/

13 US Department of Agriculture. 2021. "Winter wheat yield." https://www.nass.usda.gov/Charts_and_Maps/graphics/wwyld.pdf

14 See https://globalsolaratlas.info/

15 Zucchelli, G. et al. 2002. "The calculated in vitro and in vivo chlorophyll a absorption band shape."*Biophysical Journal* 82: 378–90; Möttus, M. et al. 2011. "Photosynthetically active radiation: Measurement and modeling." In: R. Meyers, ed. *Encyclopedia of Sustainability Science and Technology.* Berlin: Springer, 7970–8000.

16 Bathellier, C. et al. "Ribulose 1,5-bisphosphate carboxylase/oxygenase activates O_2 by electron transfer." *Proceedings of the National Academy of Sciences* 117(39).

17 Amthor, J.S. and D.D. Baldocchi. 2001. "Terrestrial higher plant respiration and net primary production." In: Roy, J., B. Saugier, and H.A. Mooney, eds., *Terrestrial Global Productivity*. San Diego: Academic Press, 33–59.
18 Zhu, X. et al. 2008. "What is the maximum efficiency with which photosynthesis can convert solar energy into biomass?" *Current Opinion in Biotechnology* 19: 153–9.
19 Bassham, J.A. and M. Calvin. 1957. *The Path of Carbon in Photosynthesis*. Engelwood Cliffs, NJ: Prentice-Hall; Calvin, M. 1989. "Forty years of photosynthesis and related activities." *Photosynthesis Research* 211: 3–16.
20 Nickell, L.G. 1993. "A tribute to Hugo P. Kortschak: The man, the scientist and the discoverer of C_4 photosynthesis." *Photosynthesis Research*. 35: 201–4.
21 Hatch, M.D. 1992. "C_4 photosynthesis: an unlikely process full of surprises." *Plant Cell Physiology* 4: 333–42.
22 Donald, C.M. and J. Hamblin. 1976. "The biological yield and harvest index of cereals as agronomic and plant breeding criteria." *Advances in Agronomy* 28: 361–405; Smil, V. 1999. "Crop residues: Agriculture's largest harvest." *BioScience* 49: 299–308.
23 Obviously, such tall crops were highly susceptible to lodging caused by strong winds and rains in thunderstorms. Agriculture and Horticulture Development Board. 2022. *An introduction to lodging in cereals*. https://ahdb.org.uk/knowledge-library/an-introduction-to-lodging-in-cereals
24 Lumpkin, T.A. 2015. "How a Gene from Japan Revolutionized the World of Wheat: CIMMYT's Quest for Combining Genes to Mitigate Threats to Global Food Security." In: Y. Ogihara et al., eds., *Advances in Wheat Genetics: From Genome to Field*. Berlin: Springer-Verlag, 13–20.
25 Thiyam-Holländer, U. et al. 2012. *Canola and Rapeseed Production, Processing, Food Quality, and Nutrition*. London: Routledge.
26 International Rice Research Institute. 2022. "Milling." http://www.knowledgebank.irri.org/step-by-step-production/postharvest/milling

27 Miracle, M.P. 1965. "The introduction and spread of maize in Africa." *Journal of African History* 6: 39–55; Ekpa, O. et al. 2019. "Sub-Saharan African maize-based foods: Processing practices, challenges and opportunities." *Food Reviews International* 35: 609–39.

28 Clark, C.M. et al. 2022. "Ethanol production in the United States: The roles of policy, price, and demand." *Energy Policy* 161: 2713; Rossi, L.M. et al. 2021. "Ethanol from sugarcane and the Brazilian biomass-based energy and chemicals sector." *Sustainable Chemical Engineering* 9: 4293–5.

29 Agricultural Marketing Resource Center. 2021. "Sweet corn." https://www.agmrc.org/commodities-products/vegetables/sweet-corn

30 Cursi, D.E. et al. 2022. "History and current status of sugarcane breeding, germplasm development and molecular genetics in Brazil." *Sugar Technology* 24: 112–33.

31 Buckley, T.N. 2019. "How do stomata respond to water status?" *New Phytologist* 224: 21–36.

32 Hatfield, J.L. and C. Dold, 2019. "Water-use efficiency: Advances and challenges in a changing climate." *Frontiers in Plant Science* 10.

33 Grossiord, C. et al. 2020. "Plant response to rising vapor pressure deficit." *New Phytologist* 226: 1550–66.

34 Briggs, L.J. and H.L. Shantz. 1913. *The Water Requirements of Plants. I. Investigation in the Great Plains in 1910 and 1911*. Washington, DC: Bureau of Plant Industry.

35 Sadras, V.O. et al. 2012. *Status of Water Use Efficiency of Main Crops*. Rome: FAO.

36 Mekonnen, M.M. and A.Y. Hoekstra. 2011. "The green, blue and grey water footprint of crops and derived crop products." *Hydrology and Earth System Sciences* 15: 1577–600.

37 Nuts are particularly water-intensive: shelled walnuts are more than 9,000 t/t and almonds about 16,000 t/t: Mekonnen, M.M. and A.Y. Hoekstra. 2010. *The Green, Blue and Grey Water Footprint of Crops and Derived Crop Products*. Volume 1: Main Report. Enschede: University of Twente.

38 Farquhar, G.D. 1997. "Carbon dioxide and vegetation." *Science* 278: 1411.

39 Kimball, B.A. 2016. "Crop responses to elevated CO_2 and interactions with H_2O, N, and temperature." *Current Opinion in Plant Biology* 31: 36–43; Basso, B. et al. 2018. "Soil Organic Carbon and Nitrogen Feedbacks on Crop Yields under Climate Change." *Agricultural and Environmental Letters.*

40 DutchGreenhouses. 2022. "Dutch Greenhouses." https://dutchgreenhouses.com/en

41 Smil, V. 2001. *Enriching the Earth: Fritz Haber, Carl Bosch, and the Transformation of World Food Production.* Cambridge, MA: MIT Press; Mosier, A. et al., eds. 2004. *Agriculture and the Nitrogen Cycle: Assessing the Impacts of Fertilizer Use on Food Production and the Environment.* Washington, DC: Island Press.

42 Heard, J. and D. Hay. 2006. *Nutrient Content, Uptake Pattern and Carbon: Nitrogen Ratios of Prairie Crops.* https://umanitoba.ca/agricultural-food-sciences/school-agriculture/school-manitoba-agronomists-conference

43 Cameron, K.C. et al. 2013. "Nitrogen losses from the soil/plant system: a review." *Annals of Applied Biology* 162: 145–73; Anas, M. et al. 2020. "Fate of nitrogen in agriculture and environment: agronomic, ecophysiological and molecular approaches to improve nitrogen use efficiency." *Biological Research* 53: 47.

44 EU Nitrogen Expert Panel. 2015. *Nitrogen Use Efficiency (NUE) – An Indicator for the Utilization of Nitrogen in Agriculture and Food Systems.* Wageningen: Wageningen University.

45 Sadras, O. and D.F. Calderini. 2021. *Crop Physiology Case Histories for Major Crops.* Amsterdam: Elsevier.

46 Liu, J. et al. 2010. "A high-resolution assessment on global nitrogen flows in cropland." *Proceedings of the National Academy of Sciences* 107: 8035–40.

47 Conant, R.T. et al. 2013. "Patterns and trends in nitrogen use and nitrogen recovery efficiency in world agriculture." *Global Biogeochemical Cycles* 27: 558–66.

48 Lassaletta, L. et al. 2014. "50-year trends in nitrogen use efficiency of world cropping systems: the relationship between yield and nitrogen input to cropland." *Environmental Research Letters* 9: 105011.

49 Kuosmanen, N. 2014. "Estimating stocks and flows of nitrogen: Application of dynamic nutrient balance to European agriculture." *Ecological Economics* 108: 68–78.

4. Why Do We Eat Some Animals and Not Others?

1 In the US, 5 percent of people (4 percent of men and 6 percent of women) identify as vegetarians: Hrynowski, Z. 2019. "What percentage of Americans are vegetarian?" https://news.gallup.com/poll/267074/percentage-americans-vegetarian.aspx. The UK share is very similar, while in France only 2.2 percent of people follow a meatless diet: https://www.vegecantines.fr/influenceurs-cantines-agir-militer-comprendre-chiffres-ressources-journalistes/chiffres-clefs-menus-vege-a-la-cantine

2 Troy, C.S. et al. 2001. "Genetic evidence for Near-Eastern origins of European cattle." *Nature* 410: 1088–91; Zeder, M.A. 2008. "Domestication and early agriculture in the Mediterranean Basin: Origins, diffusion, and impact." *Proceedings of the National Academy of Sciences* 105: 11597–604.

3 Jansen, T. et al. 2002. "Mitochondrial DNA and the origins of the domestic horse." *Proceedings of the National Academy of Sciences* 99: 10905–10.

4 Burgin, C.J. et al. 2018. "How many species of mammals are there?" *Journal of Mammalogy* 99: 1–14; Callaghan, C.T. et al. 2021. "Global abundance estimates for 9,700 bird species." *Proceedings of the National Academy of Sciences* 2021: e2023170118.

5 FAO. 2022. "Crops and livestock products." https://www.fao.org/faostat/en/#data/QCL

6 China has the largest duck and geese stocks (almost 700 million and more than 300 million birds, respectively); the US has nearly a quarter billion turkeys.

7 Ducks consume a significant share of animal foods. Different wild duck species eat, besides grasses and aquatic plants (roots and stems of bulrushes), earthworms, snails, amphibians (tadpoles, frogs), many

kinds of insects including mosquito larvae and mayfly nymphs, small crustaceans, and small fish and fish eggs.

8   Hui, D. 2012. "Food web: Concept and applications." *Nature Education Knowledge* 3(12): 6.
9   Clauss, M. 2019. "No evidence for different metabolism in domestic mammals." *Nature Ecology and Evolution* 3: 322.
10 Faas, P. 1994. *Around the Roman Table: Food and Feasting in Ancient Rome*. Chicago: University of Chicago Press, 290–1.
11 Goldstein, D.J. 2010. "The delicacy of raising and eating guinea pig." In: Haines, H.R. and C.A. Sammells, eds., *Adventures in Eating: Anthropological Experiences in Dining from around the World*. Boulder, CO: University Press of Colorado, 59–77.
12 There are also proposals for producing rabbit meat by large-scale grazing: Carangelo, N. 2019. *Raising Pastured Rabbits for Meat*. Chelsea, VT: Chelsea Green Publishing.
13 Chessa, B. et al. 2009. "Revealing the history of sheep domestication using retrovirus integrations." *Science* 324: 532–6; Alberto, F.J. et al. 2018. "Convergent genomic signatures of domestication in sheep and goats." *Nature Communications* 9: 813.
14 Landsberg, G.M. and S. Denenberg. 2016. "Social behavior of sheep." *Merck Vet Manual*. https://www.merckvetmanual.com/behavior/normal-social-behavior-and-behavioral-problems-of-domestic-animals/social-behavior-of-sheep
15 Fessler, D.M. and C.D. Navarette. 2009. "Meat is good to taboo: Dietary proscriptions as a product of the interaction of psychological mechanisms and social processes." *Journal of Cognition and Culture* 3: 1–40; Contreras, J. 2008. "Meat consumption throughout history and across cultures." *Consommer mediterràneen*. Europeo. Dossier EMS.97.004.
16 Muscle share decreases and fat share rises as pigs grow from 45 to 135 kilograms in both lean- and fat-type pigs: at 45 kilograms, fat-type carcasses are nearly 14 percent fat (lean ones only about 10 percent); at 135 kilograms the respective shares are about 44 percent and 30 percent: Lonergan, S.M. et al. 2019. "Growth curves and growth patterns."

In: Lonergan, S.M. et al., *The Science of Animal Growth and Meat Technology*. Amsterdam: Elsevier, 71–109.

17 Sowell, B.F. et al. 1999. "Social behavior of grazing beef cattle: Implications for management." *Proceedings of the American Society of Animal Science*: 1–6.

18 Italian Chianina beef animals (height of 2 meters and mass of more than 1.7 tons) are the tallest and the heaviest breed, with South Devon cattle (maxima of more than 1.6 tons) kept for both milk and meat close behind: AgronoMag. 2022. "Top 10 biggest cows in the world – largest cow breeds." https://agronomag.com/biggest-cows-world/. At the other extreme are Indian Gyr cows, weighing less than 300 kilograms at first calving.

19 Smil, V. 2017. *Energy and Civilization: A History*. Cambridge, MA: MIT Press, 87–110.

20 Liebowitz, J.J. 1992. "The persistence of draft oxen in Western agriculture." *Material Culture Review* 36(1). https://journals.lib.unb.ca/index.php/MCR/article/view/17512

21 Moore, J.H. 1961. "The ox in the Middle Ages." *Agricultural History* 35: 93.

22 Faas, P. 1994. *Around the Roman Table: Food and Feasting in Ancient Rome*. Chicago: University of Chicago Press; Eden, F.M. 1797. *The State of the Poor*. London: J. Davis.

23 Smil, V. and K. Kobayashi. 2011. *Japan's Dietary Transition and Its Impacts*. Cambridge, MA: MIT Press.

24 Rogin, L. 1931. *The Introduction of Farm Machinery*. Berkeley: University of California Press.

25 Smil, V. 2017. *Energy and Civilization: A History*. Cambridge, MA: MIT Press, 111.

26 Specht, J. 2019. *Red Meat Republic: A Hoof-to-Table History of How Beef Changed America*. Princeton, NJ: Princeton University Press.

27 US Department of Agriculture. 2022. "Livestock and Meat Domestic Data." https://www.ers.usda.gov/data-products/livestock-and-meat-domestic-data/

28 McKay, H. 2021. "Mega farms called CAFOs dominate animal agriculture industry." https://sentientmedia.org/cafo

29 The world's largest cattle feeding company has 11 feedyards in six US states, and a one-time feeding capacity of more than 985,000 animals: Five Rivers. 2022. "Cattle Feeding." https://www.fiveriverscattle.com/pages/

30 Compassion in World Farming. 2022. "About chickens farmed for meat." https://www.ciwf.org.uk/farm-animals/chickens/meat-chickens/

31 McKay, H. 2021. "Mega farms called CAFOs dominate animal agriculture industry."

32 These data are available on an annual basis in *Agricultural Statistics*, published by the US Department of Agriculture; the latest version is available at https://downloads.usda.library.cornell.edu/usda-esmis/files/j3860694x/z890sn81j/cv43pq78m/Ag_Stats_2020_Complete_Publication.pdf

33 Smil, V. 2013. *Should We Eat Meat? Evolution and Consequences of Modern Carnivory.* Chichester: John Wiley, 109–11.

34 Mottet, A. et al. 2017. "Livestock: On our plates or eating at our table? A new analysis of the feed/food debate." *Global Food Security* 14: 1–8.

35 Bonhommeau, S. et al. "Eating up the world's food web and the human trophic level." *Proceedings of the National Academy of Sciences* 110: 20617–20.

36 FAO. 2022. "Global Livestock Environmental Assessment Model (GLEAM)." https://www.fao.org/gleam/en/

37 Mekonnen, M.M. and A.Y. Hoekstra. 2010. *The Green, Blue and Grey Water Footprint of Farm Animals and Animal Products.* Enschede: University of Twente.

38 Farm Transparency Project. 2022. "Age of animals slaughtered." https://www.farmtransparency.org/kb/food/abattoirs/age-animals-slaughtered

39 Dennis, E. 2021. *Forage Production, Beef Cows and Stocking Density and Their Implications for Partial Herd Liquidation Due to Drought.* https://beef.unl.edu/beefwatch/2021/forage-production-beef-cows-and-stocking-density-and-their-implications-partial-herd

40 Dillon, J.A. et al. 2021. "Current state of enteric methane and the carbon footprint of beef and dairy cattle in the United States." *Animal Frontiers* 11: 57–68.

41 FAO. 2017. "Global Livestock Environmental Assessment Model (GLEAM)."
42 FAO. 2022. *The State of World Fisheries and Aquaculture*. Rome: FAO. https://www.fao.org/fishery/en/statistics/global-production/query/en
43 Fry, J.P. et al. 2018. "Feed conversion efficiency in aquaculture: do we measure it correctly?" *Environmental Research Letters* 13: 02401.
44 Tacon, A.G.J. and M. Metian. 2008. "Global overview on the use of fish meal and fish oil in industrially compounded aquafeeds: trends and future prospects." *Aquaculture* 285: 146–58.
45 Jackson, A. 2009. "Fish in–fish out ratios explained." *Aquaculture in Europe* 34(3): 5–10. http:// iffo.net.769soon2b.co.uk/downloads/100.pdf
46 European Commission. 2021. *Fishmeal and Fish Oil*. https://eumofa.eu/documents/20178/432372/Fishmeal+and+fish+oil.pdf?
47 Kok, B. et al. 2020. "Fish as feed: Using economic allocation to quantify the Fish In: Fish Out ratio of major fed aquaculture species." *Aquaculture* 528: 735474.
48 Auchterlonie, N. 2019. "Fish In–Fish Out Ratios." https://effop.org/wp-content/uploads/2019/10/5-IFFO-EUFM-FIFO-251019.pdf
49 Maruha Nichiro. 2022. "Fish farming." https://effop.org/wp-content/uploads/2019/10/5-IFFO-EUFM-FIFO-251019.pdf; Waycott, B. 2020. "Japan's quest to conquer bluefin farming." https://www.hatcheryinternational.com/japans-quest-to-conquer-bluefin-farming/
50 Benetti, D.D. et al., eds. 2016. *Advances in Tuna Aquaculture*. Amsterdam: Elsevier.
51 AquaBounty. 2022. "A Better Way to Raise Atlantic Salmon." https://aquabounty.com
52 Nardi, G. et al. 2021. "Atlantic cod aquaculture: Boom, bust, and rebirth?" *Journal of World Aquaculture Society* 2: 672–90.
53 Zangwill, N. 2022. "Why you should eat meat." https://aeon.co/essays/if-you-care-about-animals-it-is-your-moral-duty-to-eat-them
54 Francione, G. 2022. "We must not own animals." https://aeon.co/essays/why-morality-requires-veganism-the-case-against-owning-animals

55 Dobzhansky, T. 1973. "Nothing in biology makes sense except in the light of evolution." *American Biology Teacher* 35(3): 125–9.

5. What's More Important: Food or Smartphones?

1. World Bank. 2022. "GDP growth (annual %)." https://data.worldbank.org/indicator/NY.GDP.MKTP.KD.ZG
2. World Bank. 2022. "Agriculture, forestry, and fishing, value added (% of GDP)." https://data.worldbank.org/indicator/NV.AGR.TOTL.ZS
3. World Bank. 2022. "GFP (current US$) World." https://data.worldbank.org/indicator/NY.GDP.MKTP.CD?locations=1W
4. Global smartphone market: https://www.marketdataforecast.com/market-reports/smartphone-market. Wheat and rice prices: https://www.indexmundi.com
5. For the practices and skills of early iron metallurgy see: Smil, V. 2016. *Still the Iron Age*. Oxford: Butterworth-Heinemann, 1–17.
6. University of Southampton. 2014. "Roman amphorae: a digital resource." https://archaeologydataservice.ac.uk/archives/view/amphora_ahrb_2005/info_intro.cfm
7. On the ubiquity of *thermopolia* see: Pompeii. 2022. "Thermopolium." http://pompeiisites.org/en/archaeological-site/thermopolium. And Rome's best-preserved non-imperial tomb belongs to Marcus Vergilius Eurysaces, the owner of a large bakery: Petersen, L.H. 2003. "The baker, his tomb, his wife, and her breadbasket: The monument of Eurysaces in Rome." *Art Bulletin* 85: 230–57.
8. World Trade Organization. 2022. *World Trade Statistical Review 2021*. https://www.wto.org/english/res_e/statis_e/wts2021_e/wts2021_e.pdf
9. Bell, B. 2020. *Farm Machinery*. Aldwick: Old Pond Books; Chen. G. 2018. *Advances in Agricultural Machinery and Technologies*. London: Routledge. For the variety offered by America's largest producer of agricultural equipment see: https://www.deere.com/en/agriculture/

10 Effective waste management is a key precondition for operating such a highly concentrated source of organic waste. For what it means in poultry houses see: Ross. 2010. *Environmental Management in the Broiler House.*

11 Miller, D. 2021. "Machinery Link." https://www.dtnpf.com/agriculture/web/ag/blogs/machinerylink/blog-post/2021/01/22/machinery-industry-sees-growth-2020

12 US Department of Agriculture. 2021. *Farm Production Expenditures 2020 Summary.* https://www.nass.usda.gov/Publications/Todays_Reports/reports/fpex0721.pdf

13 US Department of Agriculture. 2022. "Ag and food sectors and the economy." https://www.ers.usda.gov/data-products/ag-and-food-statistics-charting-the-essentials/ag-and-food-sectors-and-the-economy/

14 The latest Chinese data for sectoral employment are available at: https://www.ceicdata.com/en/china/no-of-employee-by-industry-monthly/no-of-employee-agricultural-sideline-food-processing

15 Engel, E. 1857. *Die Productions- und Consumtionsverhältnisse des Königreichs Sachsen. Zeitschrift des statistischen Bureaus des Königlich Sächsischen Ministerium des Inneren* 8–9: 28–9.

16 US Department of Agriculture. 2022. "Food prices and spending." https://www.ers.usda.gov/data-products/ag-and-food-statistics-charting-the-essentials/food-prices-and-spending; Eurostat. 2020. "How much are households spending on food?" https://ec.europa.eu/eurostat/web/products-eurostat-news/-/ddn-20201228-1

17 Van Nieuwkoop, M. 2019. "Do the costs of the global food system outweigh its monetary value?" *World Bank Blogs,* June 17, 2019. https://blogs.worldbank.org/voices/do-costs-global-food-system-outweigh-its-monetary-value

18 International Energy Agency. 2021. *World Energy Balances.* Paris: IEA. https://www.iea.org/reports/world-energy-balances-overview

19 Hitaj, C. and S. Suttles. 2016. *Trends in US Agriculture's Consumption and Production of Energy: Renewable Power, Shale Energy, and Cellulosic Biomass.* Washington, DC: Department of Agriculture, Economic

Research Service. https://www.ers.usda.gov/publications/pub-details/?pubid=74661

20 Canning, P. et al. 2010. *Energy Use in the US Food System.* https://www.ers.usda.gov/webdocs/publications/46375/8144_err94_1_.pdf?v=2360
21 USDA. 2022. "Food prices and spending." https://www.ers.usda.gov/data-products/ag-and-food-statistics-charting-the-essentials/food-prices-and-spending
22 Environmental Protection Agency. 2021. *Advancing Sustainable Materials Management: 2018 Fact Sheet Assessing Trends in Materials Generation and Management in the United States.* https://www.epa.gov/sites/default/files/2021-01/documents/2018_ff_fact_sheet_dec_2020_fnl_508.pdf
23 National Bureau of Statistics. 2022. *2021 China Statistical Yearbook.* Beijing: National Statistics Press.
24 Smil, V. 2008. *Energy in Nature and Society.* Cambridge, MA: MIT Press, 291–306.
25 FAO. 2011. *Energy-smart Food for People and Climate.* Rome: FAO.
26 Global harvests from FAOSTAT. 2022. "Crops and livestock products." Energy cost of freight: assuming 3 MJ/ton-kilometer (tkm) for trucking, 1 MJ/tkm for shipping and 0.5 MJ/tkm for rail: https://www.eea.europa.eu/publications/ENVISSUEN012/page027.html
27 That translates annually to about 1.4 GJ/capita: Eurostat. 2021. "Energy consumption in households." https://ec.europa.eu/eurostat/statistics-explained/index.php?title=Energy_consumption_in_households; Hager, T.J. and R. Morawicki. "Energy consumption during cooking in the residential sector of developed nations: A review." *Food Policy* 40: 54–63.
28 Cooking in urban China consumes around 3 GJ/capita, in rural China about 5 GJ/capita: Zheng, X. et al. 2014. "Characteristics of residential energy consumption in China: Findings from a household survey." *Energy Policy* 75: 126–35. For India see: Eckholm, T. et al. 2010. "Determinants of household energy consumption in India." *Energy Policy* 38: 5696–707.
29 I assume means of 1.5–2 GJ in all affluent and middle-income countries, and 3–4 GJ in low-income nations.

30 Barthel, C. and T. Götz. 2012. *The overall worldwide saving potential from domestic refrigerators and freezers.* https://bigee.net/media/filer_public/2012/12/04/bigee_doc_2_refrigerators_freezers_worldwide_potential_20121130.pdf; Global Data Lab. 2022. "% Households with a Refrigerator." https://globaldatalab.org/areadata/fridge/

31 FAO. 2021. *The State of the World's Land and Water Resources for Food and Agriculture – Systems at breaking point. Synthesis report 2021.* Rome: FAO.

32 FAOSTAT. 2022. "Land, inputs and sustainability." https://www.fao.org/faostat/en/#data

33 Fowler, D. et al. 2013. "The global nitrogen cycle in the twenty-first century." *Philosophical Transactions of the Royal Society B* 368: 20130164.

34 Lepori, F. and F. Keck. 2012. "Effects of atmospheric nitrogen deposition on remote freshwater ecosystems." *Ambio* 41: 235–46; National Oceanic and Atmospheric Administration. 2021. "Larger-than-average Gulf of Mexico 'dead zone' measured." https://www.noaa.gov/news-release/larger-than-average-gulf-of-mexico-dead-zone-measured

35 Lynch, J. et al. 2021. "Agriculture's contribution to climate change and role in mitigation is distinct from predominantly fossil CO_2-emitting sectors." *Frontiers in Sustainable Food Systems* 4: 518039; Tubiello, F.N. et al. 2021. *Methods for estimating greenhouse gas emissions from food systems – Part III: energy use in fertilizer manufacturing, food processing, packaging, retail and household consumption.* Rome: FAO.

36 Crippa, M. et al. 2021. "Food systems are responsible for a third of global anthropogenic GHG emissions." *Nature Food* 2: 198–209.

37 Cunningham, E. 2021. "Cows, methane and the climate threat." https://fidelityinternational.com/editorial/article/esgenius-cows-methane-/and-the-climate-threat-e222b8-en5/; Waite, R. et al. 2022. "6 Pressing Questions About Beef and Climate Change, Answered." https://www.wri.org/insights/6-pressing-questions-about-beef-and-climate-change-answered

38 Van Nieuwkoop, M. 2019. "Do the costs of the global food system outweigh its monetary value?" *World Bank Blogs*, June 17, 2019. https://blogs.worldbank.org/voices/do-costs-global-food-system-outweigh-its-monetary-value

39 WHO. 2020. "World Obesity Day: All countries significantly off track to meet 2025 WHO targets on obesity." https://www.worldobesity.org/news/world-obesity-day-all-countries-significantly-off-track-to-meet-2025-who-targets-on-obesity

40 For attempts to value food waste see: von Massow, M. et al. 2019. "Valuing the multiple impacts of household food waste." *Frontiers in Nutrition* 6: 143; Conrad, Z. 2020. "Daily cost of consumer food wasted, inedible, and consumed in the United States, 2001–2016." *Nutrition Journal* 19: 35. Conrad concluded that in 2017 an average American spent more on wasted food than on gasoline, clothes, household heating, or property taxes.

41 Gillingham, K. 2019. "Carbon calculus." *Finance & Development* 7: 11.

42 Tegtemeier, E.M. and M.D. Duffy. 2004. "External costs of agricultural production in United States." *International Journal of Agricultural Sustainability* 2: 1–20.

43 FAO. 2011. *The State of the World's Land and Water Resources for Food and Agriculture (SOLAW) – Managing Systems at Risk*. Rome: FAO.

44 FAO. 2021. *The State of the World's Land and Water Resources for Food and Agriculture (SOLAW) – Systems at Breaking Point. Synthesis report 2021*. Rome: FAO.

45 Teixeira, M. 2018. "Deforestation in the Brazilian Amazon has reached a 10-year high." https://www.weforum.org/agenda/2018/11/deforestation-in-the-brazilian-amazon-reaches-decade-high/

46 Heilmayr, R. et al. 2020. "Brazil's Amazon Soy Moratorium reduced deforestation." *Nature Food* 1: 801–10.

47 Spracklen, D.V., and L. Garcia-Carreras. 2015. "The impact of Amazonian deforestation on Amazon basin rainfall." *Geophysical Research Letters* 42: 9546–52.

48 Dangar. S. et al. 2020. "Causes and implications of groundwater depletion in India: A review." *Journal of Hydrology* 596: 126103.

49 Bhanjaa, S.N. and A. Mukherjee. 2019. "In situ and satellite-based estimates of usable groundwater storage across India: Implications for drinking water supply and food security." *Advances in Water Resources* 126: 15–23.

50 Delang, C.O. 2018. "Heavy metal contamination of soils in China: standards, geographic distribution, and food safety considerations. A

review." *Die Erde* 4: 261–8; Sodango, T.H. et al. 2018. "Review of the spatial distribution, source and extent of heavy metal pollution of soil in China: Impacts and mitigation approaches." *Journal of Health & Pollution* 8: 53–70.

51 FAOSTAT. 2022. "Crops and livestock products."

52 Eurostat. 2022. "Agri-environmental indicator—livestock patterns." https://ec.europa.eu/eurostat/statistics-explained/index.php?title=Agri-environmental_indicator_-_livestock_patterns

53 Foreign Agriculture Service. 2021. "New Government Coalition Accord Reached in the Netherlands Country: Netherlands."

54 US Drought Monitor. 2022. "Map releases: March 31, 2022." https://droughtmonitor.unl.edu/

55 Williams, A.P. et al. 2020. "Large contribution from anthropogenic warming to a developing North American megadrought." *Science* 368: 314–18; Williams, A.P. et al. 2022. "Rapid intensification of the emerging southwestern North American megadrought in 2020–2021." *Nature Climate Change* 12: 232–4.

56 National Oceanic and Atmospheric Administration. 2023. "Spring Outlook: California drought cut by half with more relief to come." https://www.noaa.gov/news-release/spring-outlook-california-drought-cut-by-half-with-more-relief-to-come

6. What Should You Eat to Be Healthy?

1 Mozaffarian, D. et al. 2018. "History of modern nutrition science—implications for current research, dietary guidelines, and food policy." *British Medical Journal* 361. 2019 Jul 31: 11(8): 1760.

2 American guidelines are detailed in: Institute of Medicine. 2005. *Dietary Reference Intakes for Energy, Carbohydrate, Fiber, Fat, Fatty Acids, Cholesterol, Protein, and Amino Acids.* Washington, DC: The National Academies Press.

3 These calculations apply to expected averages. Well-known differences in individual metabolic rates can have surprisingly large effects.

4 FAO. 2001. *Food Balance Sheets: A Handbook.* Rome: FAO. https://www.fao.org/3/x9892e/x9892e00.htm
5 FAO. 2022. *Food Balance Sheets.* Rome: FAO. https://www.fao.org/faostat/en/#data/FBSH
6 FAO. 2021. *The State of Food Security and Nutrition in the World.* Rome: FAO.
7 World Food Programme. 2022. "Our work." https://www.wfp.org/our-work
8 Institute of Medicine. 2005. *Dietary Reference Intakes for Energy, Carbohydrate, Fiber, Fat, Fatty Acids, Cholesterol, Protein, and Amino Acids.* Washington, DC: The National Academies Press.; Seidelmann, S.B. et al. 2018. "Dietary carbohydrate intake and mortality: a prospective cohort study and meta-analysis." *Lancet Public Health* 3: e419–e428.
9 World Health Organization. 2007. *Protein and Amino Acid Requirements in Human Nutrition: Report of a Joint FAO/WHO/UNU Expert Consultation.*
10 Boye, J. et al. 2012. "Protein quality evaluation twenty years after the introduction of the protein digestibility corrected amino acid score method." *British Journal of Nutrition* 108 (S2): S183–S211.
11 Smil, V. 2020. *Grand Transitions.* New York: Oxford University Press.
12 The United States has led this dieting obsession, but the evidence of success has been elusive: the country has more overweight (body mass index 25–30) and obese (BMI>30) people than any other populous nation: The World Obesity Federation. 2022. *World Obesity Atlas 2022.* http://s3-eu-west-1.amazonaws.com/wof-files/World_Obesity_Atlas_2022.pdf
13 Moreno, L. et al. 2022. "Perspective: Striking a balance between planetary and human health—Is there a path forward?" *Advances in Nutrition* 13: 355–75.
14 These shares are based on FAO's food balances: actual consumption shares (adjusted for wholesale, retail, and household waste) would be 20–30 percent lower.

15 Even when ignoring their impact on the generation of greenhouse gases, just doubling the recent output of beef and other ruminant meats would be difficult to sustain.
16 As explained in the first chapter, even unimproved varieties yield enough sugar per hectare to feed about 20 people for the entire year; for modern varieties it would be commonly twice as many.
17 Phillips, R.D. 1993. "Starchy legumes in human nutrition, health and culture." *Plant Foods and Human Nutrition* 44: 195–211.
18 Polak, R. et al. 2015. "Legumes: Health benefits and culinary approaches to increase intake." *Clinical Diabetes* 33(4): 198–205.
19 Perhaps the most important antinutritional factors in legumes are trypsin inhibitors. They reduce the digestion and absorption of dietary proteins, but their deactivation causes loss of nutrients, affects their functional properties, and requires high inputs of energy: Avilés-Gaxiola, S. et al. 2018. "Inactivation methods of trypsin inhibitor in legumes: A review." *Journal of Food Science* 83: 17–29.
20 Alves, R. 2021. "The economics behind feijoada, Brazil's signature dish."
21 Selvi, A. and N. Das. 2020. *Fermented Soybean Food Products as Sources of Protein-rich Diet: An Overview*. Boca Raton, FL: CRC Press.
22 Pavan, K. et al. 2017. "Meat analogues: Health promising sustainable meat substitutes." *Critical Reviews in Food Science and Nutrition* 57(5): 923–32.
23 van Vliet, S. et al. 2020. "Plant-based meats, human health, and climate change." *Frontiers in Sustainable Food Systems* 4: 128.
24 Grandview Research. 2019. *Tofu Market Size, Share & Trends Analysis Report By Distribution Channel (Supermarkets & Hypermarkets, Grocery Stores, Online, Specialty Stores), By Region, And Segment Forecasts, 2019–2025*. https://www.grandviewresearch.com/industry-analysis/tofu-market; Otsuka Pharmaceutical. 2022. "Soybean consumption." https://www.otsuka.co.jp/en/nutraceutical/about/soylution/encyclopedia/consumption.html
25 Richter, F. 2022. "Meat Substitutes Still a Tiny Sliver of US Meat Market." https://www.statista.com/chart/26695/meat-substitute-sales-in-the-us/

26 Tuomisto, H.L. et al. 2017. "Effects of environmental change on population nutrition and health: A comprehensive framework with a focus on fruits and vegetables." *Wellcome Open Research* 2: 21.

27 Center for Urban Education about Sustainable Agriculture. 2022. "Seasonality charts: Vegetables." https://foodwise.org/eat-seasonally/seasonality-charts/

28 Mekonnen, M.M. and A.Y. Hoekstra. 2010. *The Green, Blue and Grey Water Footprint of Crops and Derived Crop Products*. Volume 1: Main Report. Enschede: University of Twente.

29 Neira, D.P. et al. 2018. "Energy use and carbon footprint of the tomato production in heated multi-tunnel greenhouses in Almeria within an exporting agri-food system context." *Science of the Total Environment* 628: 1627–36.

30 Cabbage has about three times as much vitamin C as a fresh tomato and, unlike tomatoes, it can be stored for weeks after harvesting. It is also easily transformed into sauerkraut (fermented by lactic acid bacteria), which preserves nearly half of the original vitamin C content.

31 Mekonnen, M.M. and A.Y. Hoekstra. 2010. *The Green, Blue and Grey Water Footprint of Crops and Derived Crop Products*. Volume 1: Main Report. Enschede: University of Twente.

32 Hullings, A.G. et al. 2020. "Whole grain and dietary fiber intake and risk of colorectal cancer in the NIH-AARP Diet and Health Study cohort." *The American Journal of Clinical Nutrition* 112: 603–12.

33 Kahleova, D. et al. 2018. "A plant-based high-carbohydrate, low-fat diet in overweight individuals in a 16-week randomized clinical trial: The role of carbohydrates." *Nutrients* 10: 1302.

34 Khan, S.A. et al. 2021. "Effect of omega-3 fatty acids on cardiovascular outcomes: A systematic review and meta-analysis." *EClinicalMedicine*.

35 Amini, M. et al. 2021. "Trend analysis of cardiovascular disease mortality, incidence, and mortality-to-incidence ratio: results from global burden of disease study 2017." *BMC Public Health* 21: 401.

36 Mahmood, S.S. et al. 2014. "The Framingham Heart Study and the epidemiology of cardiovascular diseases: A historical perspective." *Lancet* 383: 999–1008.

37 Keys, A. 1980. *Seven Countries: A Multivariate Analysis of Death and Coronary Heart Disease.* Cambridge, MA: Harvard University Press; Keys, A. and M. Keys. 1975. *How to Eat Well and Stay Well the Mediterranean Way.* New York: Doubleday.

38 Liquid fats are turned into solids by the addition of hydrogen, in order to lengthen the shelf life of processed foods and to reduce production costs.

39 Ferrières, J. 2004. "The French paradox: lessons for other countries." *Heart* 90: 107–11; Renaud, S. and M. de Lorgeril. 1992. "Wine, alcohol, platelets, and the French paradox for coronary heart disease." *Lancet* 339: 1523–6.

40 Chowdhury, R. et al. 2014. "Association of dietary, circulating, and supplement fatty acids with coronary risk: a systematic review and meta-analysis." *Annals of Internal Medicine* 160: 398–406; De Souza, R.J. et al. 2015. "Intake of saturated and trans unsaturated fatty acids and risk of all cause mortality, cardiovascular disease, and type 2 diabetes: systematic review and meta-analysis of observational studies." *British Medical Journal.*

41 American Heart Association. 2020. "Dietary Cholesterol and Cardiovascular Risk. A Science Advisory from the American Heart Association." *Circulation* 141: e39–e53.

42 Bier, D.M. 2016. "Saturated fats and cardiovascular disease: Interpretations not as simple as they once were." *Critical Reviews in Food Science and Nutrition* 56(12): 1943–6; Siri-Tarino, P.W. et al. 2010. "Saturated fat, carbohydrate, and cardiovascular disease." *American Journal of Clinical Nutrition* 91: 502–9.

43 Seidelmann, S.B. et al. 2018. "Dietary carbohydrate intake and mortality: a prospective cohort study and meta-analysis." *Lancet Public Health* 3: e419–e428.

44 Hu is cited in: O'Connor, A. 2014. "Study doubts saturated fat's link to heart disease." *New York Times*, March 18, page A3.

45 For example, in Europe. Finland and Greece are far apart both spatially and nutritionally but their life expectancies (79.2 and 79.5 for males, 84.5 for both countries for females) are virtually identical.

46 Eurostat. 2022. "Life expectancy across EU regions in 2020." https://ec.europa.eu/eurostat/web/products-eurostat-news/-/ddn-20220427-1

47 And Spain was not alone; the Mediterranean diet became less so also in France, Italy, Croatia, and Greece: Smil, V. 2016. "Addio to the Mediterranean diet." *IEEE Spectrum*, September 2016: 24.

48 FAO. 2022. Food balance sheets; Landgeist. 2021. "Meat consumption." https://landgeist.com/2021/10/05/meat-consumption-in-europe/

49 Serra-Majem, L. et al. 1995. "How could changes in diet explain changes in coronary heart disease mortality in Spain—The Spanish Paradox." *American Journal of Clinical Nutrition* 61: S1351-S1359.

50 Cayuela, L. et al. 2021. "Is the pace of decline in cardiovascular mortality decelerating in Spain?" *Revista Española de Cardiología* 74: 750–67.

51 Xiao, H. et al. 2015. "The puzzle of the missing meat: Food away from home and China's meat statistics." *Journal of Integrative Agriculture* 14(6): 1033–44.

52 Chen, H. et al. 2018. "Understanding the rapid increase in life expectancy in Shanghai, China: a population-based retrospective analysis." *BMC Public Health* 18: 256.

53 A recent study has also shown that dietary supplements have only a minimal role in weight loss attempts: Batsis, J.A. et al. 2021. "A systematic review of dietary supplements and alternative therapies for weight loss." *Obesity* 29(7): 1102–13.

54 Olson, R. et al. 2021. "Food fortification: The advantages, disadvantages and lessons from *Sight and Life* programs." *Nutrients* 13: 1118.

55 Johns Hopkins University. 2019. *Methodology Report: Decade of Vaccines Economics (DOVE) Return on Investment Analysis*. https://static1.squarespace.com/static/556deb8ee4b08a534b8360e7/t/5d56d54c6dae8d00014ef72d/1565971791774/DOVE-ROI+Methodology+Report+16AUG19.pd

56 Bailey, R.L. et al. 2015. "The epidemiology of global micronutrient deficiencies." *Annals of Nutrition and Metabolism* 66 (suppl. 2): 22–33; Bhutta, Z.A. et al. 2013. "Meeting the challenges of micronutrient

malnutrition in the developing world." *British Medical Bulletin* 106: 7–17.
57 Wirth, J.P. et al. 2017. "Vitamin A supplementation programs and country-level evidence of vitamin A deficiency." *Nutrients* 9: 190.
58 And in Africa only a handful of countries, including the Republic of South Africa, Kenya, and Uganda, had data collected since 2010.
59 Safri, S. et al. 2021. "Burden of anemia and its underlying causes in 204 countries and territories, 1990–2019: results from the Global Burden of Disease Study 2019." *Journal of Hematological Oncology* 14: 185.
60 Miller, J.L. 2013. "Iron deficiency anemia: A common and curable disease." *Cold Spring Harbor Perspectives in Medicine* 3: a011866/.
61 Zhu, X. et al. 2020. "Correlates of nonanemic iron deficiency in restless legs syndrome." *Frontiers in Neurology*, 30 April 2020.
62 Mei, Z. et al. 2021. "Physiologically based serum ferritin thresholds for iron deficiency in children and non-pregnant women: a US National Health and Nutrition Examination Surveys (NHANES) serial cross-sectional study." *Lancet Haematology* 8(8): e572–e582.
63 Biban, B.G. and C. Lichiaropol. 2017. "Iodine deficiency, still a global problem?" *Current Health Sciences Journal* 43: 103–11.
64 Rah, J.R. et al. 2015. "Towards universal salt iodisation in India: achievements, challenges and future actions." *Maternal and Child Nutrition* 11: 483–96.
65 Gupta, S. et al. 2020. "Zinc deficiency in low- and middle-income countries: prevalence and approaches for mitigation." *Journal of Human Nutrition and Dietetics* 33: 624–43.
66 Caulfield, L.E. and R.E. Black. 2003. "Zinc deficiency. Comparative quantification of health risks: global and regional burden of disease attributable to selected major risk factors." *World Health Organization* 1: 257–80; Kumssa, D.B. et al. 2015. "Dietary calcium and zinc deficiency risks are decreasing but remain prevalent." *Scientific Reports* 5: 10974–84; Wessells, K.R. and K.H. Brown. 2012. "Estimating the global prevalence of zinc deficiency: Results based on zinc availability in national food supplies and the prevalence of stunting." *PLoS ONE* 7: 1–11.

67 Belay, A. et al. 2021. "Zinc deficiency is highly prevalent and spatially dependent over short distances in Ethiopia." *Scientific Reports* 11: 6510.
68 Kashi, B. et al. 2019. "Multiple micronutrient supplements are more cost-effective than iron and folic acid: Modeling results from 3 high-burden Asian countries." *Journal of Nutrition* 149: 1222–9.
69 FAO. 2021. *The State of Food Security and Nutrition in the World*. Rome: FAO.
70 Tarasuk, V. and A. Mitchell. 2020. *Household Food Insecurity in Canada, 2017–18*. "Toronto: Research to identify policy options to reduce food insecurity (PROOF)." https://proof.utoronto.ca
71 US Department of Agriculture. 2022. *Methodology Report: Decade of Vaccine Economics*. Supplemental Nutrition Assistance Program (SNAP); Kim, L. 2022. "France considers giving out food subsidies amid rising prices." *Forbes*, March 22, 2022; Josling, T. 2011. *Global Food Stamps: An Idea Worth Considering?* Geneva: International Centre for Trade and Sustainable Development.
72 *Global Yield Gap Atlas*. 2022. Lincoln, NE: University of Nebraska. https://www.yieldgap.org
73 US Department of Agriculture. 2021. *Nigeria: Grain and Feed Update*.
74 The International Crops Research Institute for the Semi-Arid Tropics. 2022. "Groundnut pyramids in Nigeria: Can they be revived?"
75 Ji, Y. et al. 2020. "Will China's fertilizer use continue to decline? Evidence from LMDI analysis based on crops, regions and fertilizer types." *PLoS ONE* 15(8): e0237234.
76 The Global Economy. 2022. "Political Stability in Sub Sahara Africa." https://www.theglobaleconomy.com/rankings/wb_political_stability/Sub-Sahara-Africa/
77 Nippon.com. 2021. "Japan's Food Self-Sufficiency Rate Matches Record Low." https://www.nippon.com/en/japan-data/h01101/
78 OECD/FAO. 2021. *OECD-FAO Agricultural Outlook 2021–2030*. Paris: OECD Publishing.
79 Hoddinott, J. 2013. "The Economic Cost of Malnutrition." https://www.nutri-facts.org/content/dam/nutrifacts/media/media-book/RTGN_chapter_05.pdf

7. Feeding a Growing Population with Reduced Environmental Impacts: Dubious Solutions

1 United Nations. 2019. *2019 Revision of World Population Prospects*. New York: UN. https://population.un.org/wpp/
2 Bricker, D. 2021. "Bye, bye, baby? Birthrates are declining globally—here's why it matters." https://www.weforum.org/agenda/2021/06/birthrates-declining-globally-why-matters/
3 FAO. 2022. "FAOSTAT—Food Balances (2010–)."
4 In 2020 and 2021, China's imports of wheat and other cereals reached the highest level since the year 2000, driven by a strong demand for animal feed.
5 FAO. 2022. "FAOSTAT—Crops and Livestock Products."
6 OECD/FAO. 2021. *OECD-FAO Agricultural Outlook 2021–2030*. Paris: OECD Publishing.
7 Ibid.
8 Vaughan, C. 2020. "Ethanol market is disturbing to American farmers." *Successful Farming*, https://www.agriculture.com/news/business/ethanol-market-is-disturbing-as-hell-to-american-farmers-and-now-there-s-covid-19; Samora, R. 2021. "Brazil 2021/22 sugar output seen down sharply; adverse weather cited." https://www.reuters.com/markets/commodities/brazil-202122-sugar-output-falls-sharply-due-adverse-weather-2021-12-16/
9 I have looked at major instances of failed, or greatly exaggerated, technical innovations in: Smil, V. 2023. *Invention and Innovation: A Brief History of Hype and Failure*. Cambridge, MA: MIT Press.
10 For critical assessments of the practice see: Kirchmann, H. and L. Bergström, eds. 2008. *Organic Crop Production—Ambitions and Limitations*. Berlin: Springer.
11 Russel, D.A. and G.W. Williams. 1977. "History of chemical fertilizer development." *Soil Science Society of America Journal*.
12 FAO. 2022. "FAOSTAT—Fertilizers by Nutrient."
13 Badgley, C. et al. 2007. "Organic agriculture and the global food supply." *Renewable Agriculture and Food Systems* 22: 86–108.

14 de Ponti, T. et al. 2012. "The crop yield gap between organic and conventional agriculture." *Agricultural Systems* 108: 1–9.
15 Alvarez, R. 2021. "Comparing productivity of organic and conventional farming systems: A quantitative review." *Archives of Agronomy and Soil Science*.
16 Bergström, L. et al. 2008. "Widespread Opinions About Organic Agriculture—Are They Supported by Scientific Evidence?" In: Kirchmann, H. and L. Bergström, eds., *Organic Crop Production—Ambitions and Limitations*. Berlin: Springer, 3.
17 Hülsbergen, K.-J. et al. 2023. *Umwelt- und Klimawirkungen des ökologischen Landbaus*. Berlin: Verlag Dr. Köster.
18 Packroff, J. 2023. "Eat less meat, we need space for biofuels, German producer says." https://www.euractiv.com/section/politics/news/eat-less-meat-we-need-space-for-biofuels-german-producer-says/
19 Estel, S. et al. 2016. "Mapping cropland-use intensity across Europe using MODIS NDVI time series." *Environmental Research Letters* 11: 024015; Jeong, S-J. et al. 2014. "Effects of double cropping on summer climate of the North China Plain and neighbouring regions." *Nature Climate Change* 4: 615–19; Waha, K. et al. 2020. "Multiple cropping systems of the world and the potential for increasing cropping intensity." *Global Environmental Change* 64: 102131.
20 Siebert, S. et al. 2010. "Global patterns of cropland use intensity." *Remote Sensing* 2: 1625–43.
21 Nadeem, F. et al. 2019. "Crop rotations, fallowing, and associated environmental benefits." *Environmental Science*. https://doi.org/10.1093/acrefore/9780199389414.013.197
22 Roesch-McNally, G.E. et al. 2017. "The trouble with cover crops: Farmers' experiences with overcoming barriers to adoption." *Renewable Agriculture and Food Systems* 33(4): 322–33.
23 Leavitt, M. and M. Smith. 2020. "Nine things that can go wrong with cover crops: prevention and management." https://alseed.com/nine-things-that-can-go-wrong-with-cover-crops-prevention-and-management

24 For details on the agronomy of leguminous cover species see: Islam, R. and B. Sherman, eds. 2021. *Cover Crops and Sustainable Agriculture*. Boca Raton, FL: CRC Press.
25 Runck, B.C. et al. 2020. "The hidden land use cost of upscaling cover crops." *Communications Biology* 3: 300.
26 Chorley, G.P.H. 1981. "The agricultural revolution in Northern Europe, 1750–1880: nitrogen, legumes, and crop productivity." *Economic History* 34: 71–93.
27 Smil, V. 2004. *China's Past, China's Future*. New York: Routledge Curzon.
28 FAO. 2022. "FAOSTAT—Livestock Manure."
29 This would result in especially high labor demands in most of sub-Saharan Africa, and in those parts of Asia where staple crop cultivation is done predominantly by smallholders.
30 Bordonal, R.d.O. et al. 2018. "Sustainability of sugarcane production in Brazil." *Agronomy for Sustainable Development* 38: article 13.
31 Wagoner, P. and J.R. Schaeffer. 1990. "Perennial grain development: Past efforts and potential for the future." *Critical Reviews in Plant Sciences* 9: 381–408; Jackson, W. 1980. *New Roots for Agriculture*. Lincoln, NE: University of Nebraska Press; Kantar, M.B. et al. 2016. "Perennial grain and oilseed crops." *Annual Review of Plant Biology* 67: 703–29.
32 Land Institute. 2022. "Transforming Agriculture, Perennially." https://landinstitute.org/our-work/perennial-crops/kernza/; Kernza. 2022. "Kernza goes to market." https://kernza.org/the-state-of-kernza/
33 Rudoy, D. et al. 2021. "Review and analysis of perennial cereal crops at different maturity stages." *IOP Conf. Series: Earth and Environmental Science* 937: 022111.
34 Shen, Y. et al. 2019. "Can ratoon cropping improve resource use efficiencies and profitability of rice in central China?" *Field Crops Research* 234: 66–72.
35 Zhang, Y. et al. 2021. "An innovated crop management scheme for a perennial rice cropping system and its impacts on sustainable rice production." *European Journal of Agronomy* 122: 126186.
36 Zhang, S. et al. 2023. "Sustained productivity and agronomic potential of perennial rice." *Nature Sustainability* 6: 28–38.

37 Shaobing Peng, Professor and Director, Crop Physiology and Production Center, Huazhong Agricultural University Wuhan, Hubei, email of April 27, 2022.
38 Cassman, K.G. and D.J. Connor. 2022. "Progress towards perennial grains for prairies and plains." *Outlook on Agriculture* 51(1).
39 McAlvay, A.C. et al. 2022. "Cereal species mixtures: an ancient practice with potential for climate resilience. A review." *Agronomy for Sustainable Development* 42: 100.
40 Loomis, R.S. 2022. "Perils of production with perennial polycultures." *Outlook on Agriculture* 51(1).
41 Pankiewicz, V.C.S. et al. 2019. "Are we there yet? The long walk towards the development of efficient symbiotic associations between nitrogen-fixing bacteria and non-leguminous crops." *BMC Biology* 17: 99; Bloch, S.E. et al. 2020. "Harnessing atmospheric nitrogen for cereal crop production." *Current Opinion in Biotechnology* 62: 181–8; Huisman, R. and R. Geurts. 2020. "A roadmap toward engineered nitrogen fixing nodule symbiosis." *Plant Communications* 1(1).
42 Giles Oldroyd, cited in: Arnason, R. 2015. "The search for the holy grail: nitrogen fixation in cereal crops." *The Western Producer*.
43 Lin, M.T. et al. 2014. "A faster Rubisco with potential to increase photosynthesis in crops." *Nature* 513: 547–50; Carmo-Silva, E. et al. 2015. "Optimizing Rubisco and its regulation for greater resource use efficiency." *Plant, Cell and Environment* 38: 1817–32.
44 Somerville, C.R. 1986. "Future prospects for genetic manipulation of Rubisco." *Philosophical Transaction of the Royal Society B* 313: 459–69.
45 Hennacy, J. and M.C. Jonikas. 2020. "Prospects for engineering biophysical CO_2 concentrating mechanisms into land plants to enhance yields." *Annual Review of Plant Biology* 71: 461–85.
46 Souza, A.P. de et al. 2022. "Soybean photosynthesis and crop yield are improved by accelerating recovery from photoprotection." *Science* 377: 851–4.
47 Sinclair, T. et al. 2023. "Soybean photosynthesis and crop yield are improved by accelerating recovery from photoprotection." *Science* 379.

48 Kupferschmidt, K. 2013. "Here it comes... The $375,000 lab-grown beef burger." *Science.* https://www.science.org/content/article/here-it-comes-375000-lab-grown-beef-burger; Ramani, S. et al. 2021. "Technical requirements for cultured meat production: a review." *Journal of Animal Science Technology* 63: 681–92; Phua, R. 2020. "Lab-grown chicken dishes to sell for S$23 at private members' club 1880 next month." https://www.channelnewsasia.com/singapore/lab-grown-chicken-nuggets-1880-eat-just-price-customers-495251

49 Good Food Institute. 2021. *Cultivated Meat and Seafood.* Washington, DC: Good Food Institute.

50 Good Food Institute. 2021. *Alternative Seafood.* Washington, DC: Good Food Institute.

51 Good Food Institute. 2021. *Deep Dive: Cultivated Meat Bioprocess Design.* Washington, DC: Good Food Institute.

52 Good Food Institute. 2022. *Cultivated Meat Scaffolding.* Washington, DC: Good Food Institute.

53 Vergeer, R. et al. 2021. *TEA of cultivated meat. Future projections of different scenarios.* Delft: CE Delft.

54 Hughes, H. 2021. *Review of Techno-Economic Assessment of Cultivated Meat.* https://www.linkedin.com/pulse/cultivated-meat-myth-reality-paul-wood-ao

55 Humbird, D. 2021. "Scale-up economics for cultured meat." *Biotechnology and Bioengineering* 118: 3239–50.

56 Wittman, C. et al., eds. 2017. *Industrial Biotechnology: Products and Processes.* Weinheim: Wiley VCH.

57 Mattick, C.S. et al. 2015. "Anticipatory life cycle analysis of in vitro biomass cultivation for cultured meat production in the United States." *Environmental Science and Technology* 49: 11941–9.

58 Belkhir, L. and A. Elmeligi. 2019. "Carbon footprint of the global pharmaceutical industry and relative impact of its major players." *Journal of Cleaner Production* 214: 185–94.

59 Tiseo, K. et al. 2020. "Global trends in antimicrobial use in food animals from 2017 to 2030." *Antibiotics* 9(12): 918; Van Boeckel, T.P. et al. 2015. "Global trends in antimicrobial use in food animals." *Proceedings of the National Academy of Sciences* 112: 5649–54.

60 Good Food Institute. 2021. *Cultivated Meat and Seafood.* Washington, DC: Good Food Institute.
61 ResearchAndMarkets. 2022. *Global Market for Cultured Meat—Market Size, Trends, Competitors, and Forecasts.* https://www.researchandmarkets.com/reports/5515331/global-market-for-cultured-meat-market-size
62 Delft University of Technology. 2022. "Dutch government confirms €60M investment into cellular agriculture." https://www.tudelft.nl/en/2022/tnw/dutch-government-confirms-eur60m-investment-into-cellular-agriculture
63 Good Food Institute. 2023. *Cultivated Meat and Seafood.* Washington, DC: Good Food Institute.
64 Plant-Based Food Association. 2023. https://members.plantbasedfoods.org/checkout/2022-summary-report
65 van Vliet, S. et al. 2020. "Plant-based meats, human health, and climate change." *Frontiers in Sustainable Food Systems* 4: 128.
66 Smil, V. 2023. *Invention and Innovation: A Brief History of Hype and Failure.* Cambridge, MA: MIT Press.

8. Feeding a Growing Population: What Would Work

1 Flach, B. and M. Selten. 2021. "Dutch Parliament approves law to reduce nitrogen emissions." *Global Agricultural Information Network* January 7, 2021.
2 Bomgardner, M.M. and B.E. Erickson. 2021. "How soil can help solve our climate problem." *Chemical and Engineering News* 99: 18.
3 Khan, S. et al. 2019. "Development of drought-tolerant transgenic wheat: Achievements and limitations." *International Journal of Molecular Sciences* 20(13): 3350.
4 Cisternas, I. et al. 2020. "Systematic literature review of implementations of precision agriculture." *Computers and Electronics in Agriculture*; Nowak, B. et al. 2021. "Precision agriculture: Where do we stand? A review of the adoption of precision agriculture technologies on field crops farms in developed countries." *Agricultural Research* 10: 515–22.
5 Here is one of many examples, from northern Ghana: Danso-Abbeam, G. et al. 2018. "Agricultural extension and its effects on farm

productivity and income: insight from Northern Ghana." *Agriculture & Food Security* 7: 74.

6 This, of course, departs from the now widely shared belief that radical transformations rather than incremental gains have become the norm. I have addressed these matters in the already-cited book on invention and innovation.

7 US Environmental Protection Agency. 2020. *Advancing Sustainable Materials Management: 2018 Fact Sheet Assessing Trends in Materials Generation and Management in the United States*. Washington, DC: US EPA.

8 Food Share. 2022. "Shelf Life guide." https://foodshare.com/wp-content/uploads/2018/06/Food-Shelf-Life-Guide.pdf

9 Even in Canada, the proposed national building code will not mandate triple-pane windows until 2030: https://www.usglassmag.com/canada-code-updates-what-you-need-to-know/

10 Less than 10 percent of annually generated plastic is recycled: Letcher, T., ed. 2020. *Plastic Waste and Recycling*. Amsterdam: Elsevier.

11 Gustavsson, J. et al. 2011. *Global Food Losses and Food Waste*. Rome: FAO.

12 FAO. 2014. *Global Initiative on Food Loss and Waste Reduction*. Rome: FAO.

13 Waste and Resources Action Programme (WRAP). 2007. *The Food We Waste*. Banbury: WRAP. https://wrap.s3.amazonaws.com/the-food-we-waste.pdf

14 WRAP. 2012. *Household Food and Drink Waste in the United Kingdom 2012*.

15 Gunders, D. 2012. *Wasted: How America Is Losing Up to 40 Percent of Its Food from Farm to Fork to Landfill*. Washington, DC: NRDC.

16 Nikkel, L. et al. 2019. *The Avoidable Crisis of Food Waste: Roadmap*. Toronto: Second Harvest and Value Chain Management International.

17 PubMed.gov. 2022. "Food waste." https://pubmed.ncbi.nlm.nih.gov/?term=food+waste

18 Bellemare, M.F. et al. 2017. "On the measurement of food waste." *American Journal of Agricultural Economics* 99: 1148–58.

19 Conrad, Z. 2020. "Daily cost of consumer food wasted, inedible, and consumed in the United States, 2000–2016." *Nutrition Journal* 19: 35.

20 Verma, M. et al. 2020. "Consumers discard a lot more food than widely believed: Estimates of global food waste using an energy gap approach and affluence elasticity of food waste." *PLoS ONE* 15(2): e0228369.
21 Lopez Barrera, E. et al. 2021. "Global food waste across the income spectrum: Implications for food prices, production and resource use." *Food Policy* 98: 101874.
22 Li, C. et al. 2022. "A systematic review of food loss and waste in China: Quantity, impacts and mediators." *Journal of Environmental Management* 303.
23 Gao, L. et al. 2021. "Vanity and food waste: Empirical evidence from China." *Journal of Consumer Affairs* 55: 1211–25.
24 Luo, Y. et al. 2021. "Household food waste in rural China: A noteworthy reality and a systematic analysis." *Waste Management & Research* 39: 1389–95.
25 The National People's Congress of the People's Republic of China. 2021. *Order of the President of the People's Republic of China No. 78*. http://www.npc.gov.cn/englishnpc/c23934/202112/f4b687aa91b0432baa4b6bdee8aa1418.shtml/
26 Liljestrand, K. 2017. "Logistics solutions for reducing food waste." *International Journal of Physical Distribution & Logistics Management* 47: 318–39.
27 Eurostat. 2022. "Daily calorie supply per capita by source."
28 AKCP. 2020. "Importance of Monitoring Food Storage Warehouse Environmental Conditions." https://www.akcp.com/blog/how-to-monitor-food-storage-warehouse-conditions/#:~:text=Importance%20of%20Monitoring%20Food%20Storage%20Warehouse%20Environmental%20Conditions
29 Neff, R.A. et al. 2015. "Wasted food: US consumers' reported awareness, attitudes, and behaviors." *PLoS ONE* 10(6): e0127881.
30 Verghese, K. et al. 2015. "Packaging's role in minimizing food loss and waste across the supply chain." *Packaging Technology and Science* 28: 603–20; Evans, D. 2011. "Blaming the consumer once again: The social and material contexts of everyday food waste practices in some English households." *Critical Public Health* 21: 429–40.

31 See, for example: Davis, J.L. 2003. "French Secrets to Staying Slim: US and French Portion Sizes Vary Vastly." https://www.webmd.com/diet/news/20030822/french-secrets-to-staying-slim

32 Stöckli, S. et al. 2018. "Normative prompts reduce consumer food waste in restaurants." *Waste Management* 77: 532–6.

33 And, as Japan (with average daily supply at just 2,700 kcal/capita) shows, without any negative effects on life expectancy!

34 Malito, A. 2017. "Grocery stores carry 40,000 more items than they did in the 1990s." *MarketWatch*. https://www.marketwatch.com/story/grocery-stores-carry-40000-more-items-than-they-did-in-the-1990s-2017-06-07

35 Houck, B. 2019. "There's too much yogurt." *Eater*, April 9, 2019. https://www.eater.com/2019/4/9/18303432/yogurt-decline-us-sales

36 Zeballos, E. and W. Sinclair. 2020. "Average Share of Income Spent on Food in the United States Remained Relatively Steady From 2000 to 2019." https://www.ers.usda.gov/amber-waves/2020/november/average-share-of-income-spent-on-food-in-the-united-states-remained-relatively-steady-from-2000-to-2019/; Eurostat. 2020. "How much are households spending on food." https://ec.europa.eu/eurostat/web/products-eurostat-news/-/ddn-20201228-1; Statistics Bureau of Japan. 2022. "Summary of the Latest Month on Family Income and Expenditure Survey." https://www.stat.go.jp/english/data/kakei/156.html; National Bureau of Statistics of China. 2022. "Households' Income and Consumption Expenditure in 2021." http://www.stats.gov.cn/english/PressRelease/202201/t20220118_1826649.html

37 Initial reactions have been largely on the catastrophic side, including the World Food Programme's 2022 *Global Report on Food Crises*: https://www.wfp.org/publications/global-report-food-crises-2022

38 For an extensive international comparison see: Regmi, A. and J.L. Seale, Jr. 2010. *Cross-Price Elasticities of Demand Across 114 Countries*. Washington, DC: USDA.

39 For post-1960 trends, see FAOSTAT. 2022. "Food balances." https://www.fao.org/faostat/en/#data/. FBSH For the latest data see: https://landgeist.com/2021/10/05/meat-consumption-in-europe/

40 Viande. 2018. "La consommation de viande diminue régulièrement." https://www.franceagrimer.fr/fam/content/download/66996/document/NCO-VIA-Consommation_viandes_France_2020.pdf?version=2
41 In 2021 the per capita average for poultry, pork and beef was about 42 kilograms: OECD. 2022. "Meat consumption."
42 Gibbs, H.K. and J.M. Salmon. 2015. "Mapping the world's degraded lands." *Applied Geography* 57: 12–21.
43 Vallentine, J.F. 1990. *Grazing Management*. San Diego: Academic Press; Smil, V. 2013. *Should We Eat Meat?* Chichester: Wiley-Blackwell.
44 Oil cakes (processing residues after extracting oil from seeds) are among the best high-protein sources of feed: FAO. 2004. *Protein Sources for the Animal Feed Industry*. Rome: FAO.
45 Brown, L. 1994. "How China could starve the world: its boom is consuming global food supplies." *Washington Post*, Outlook Section, August 24, 1994; Brown, L. 1995. *Who Will Feed China?: Wake-Up Call for a Small Planet*. New York: W.W. Norton.
46 Smil, V. 1995. "Who will feed China?" *China Quarterly* 143: 801–13.
47 National Bureau of Statistics. 2022. *2021 China Statistical Yearbook*. Beijing: National Statistics Press.
48 Monbiot, G. 2022. "Contagious collapse." https://www.monbiot.com/2022/05/20/contagious-collapse/
49 In fact, by far the largest bankruptcy in 2008 was not of a bank but of a financial services company, when Lehman Brothers, holding assets of more than $600 billion, folded on September 15, 2008—but, contrary to widely held opinion, this was not the cause of financial crisis: Skeel, D. 2018. "History credits Lehman Brothers' collapse for the 2008 financial crisis. Here's why that narrative is wrong." https://www.brookings.edu/articles/history-credits-lehman-brothers-collapse-for-the-2008-financial-crisis-heres-why-that-narrative-is-wrong/
50 Huang, J. et al. 2017. "The prospects for China's food security and imports: Will China starve the world via imports?" *Journal of Integrative Agriculture* 16: 2933–44.
51 Grimmelt, A. et al. 2023. "For love of meat: Five trends in China that meat executives must grasp." https://www.mckinsey.com/industries/

consumer-packaged-goods/our-insights/for-love-of-meat-five-trends-in-china-that-meat-executives-must-grasp

52 Deng, N. et al. 2019. "Closing yield gaps for rice self-sufficiency in China." *Nature Communications* 10: 1725.

53 Global Network Against Food Crises. 2022. *Global Report on Food Crises*. Rome: FAO.

54 ten Bergen, H.F.M. et al. 2019. "Maize crop nutrient input requirements for food security in sub-Saharan Africa." *Global Food Security* 23: 9–21.

55 Saito, K. et al. 2019. "Yield-limiting macronutrients for rice in sub-Saharan Africa." *Geoderma* 338: 546–54.

56 Setilkhumar, K. et al. 2020. "Quantifying rice yield gaps and their causes in Eastern and Southern Africa." *Journal of Agronomy and Crop Science* 206: 478–90.

57 Guilpart, N. et al. 2017. "Rooting for food security in Sub-Saharan Africa." *Environmental Research Letters* 12: 114036.

58 Harahagazwe, D. et al. 2018. "How big is the potato (*Solanum tuberosum* L.) yield gap in Sub-Saharan Africa and why? A participatory approach." *Open Agriculture* 3(2): 180–9.

59 Anago, F.N. et al. 2021. "Cultivation of cowpea. Challenges in West Africa for food security: Analysis of factors driving yield gap in Benin." *Agronomy* 11: 1139.

60 Dzanku, F.M. et al. 2015. "Yield gap-based poverty gaps in rural Sub-Saharan Africa." *World Development* 67: 336–62.

61 van Ittersum, M.K. et al. 2016. "Can sub-Saharan Africa feed itself?" *Proceedings of the National Academy of Sciences* 113: 14964–9.

62 For the intensification of the water cycle: Olmedo, E. et al. 2022. "Increasing stratification as observed by satellite sea surface salinity measurements." *Scientific Reports* 12: 6279. For the changing dates of wine grape harvests: Labbé, T. et al. 2019. "The longest homogeneous series of grape harvest dates, Beaune 1354–2018, and its significance for the understanding of past and present climate." *Climates of the Past* 15: 1485–501. And for changing nutritional content (zinc and protein deficiencies): Smith, M.R. and S.S. Myers. 2018. "Impact of anthropogenic CO_2 emissions on global human nutrition." *Nature Climate Change* 8: 834–9.

63 For the biosphere's greening see: Zhu, Z. et al. 2016. "Greening of the Earth and its drivers." *Nature Climate Change* 6: 791–5. For an example of higher crop yields: Degener, J.F. 2015. "Atmospheric CO_2 fertilization effects on biomass yields of 10 crops in northern Germany." *Environmental Science* 3. For corn yields: Rizzo, G. et al. 2022. "Climate and agronomy, not genetics, underpin recent maize yield gains in favorable environments." *Proceedings of the National Academy of Sciences* 119(4): e2113629119.

64 Knauer. J. et al. 2023. "Higher global gross primary productivity under future climate with more advanced representations of photosynthesis." *Science Advances* 9: eadh9444/.

65 Pugh, T.A.M. et al. 2016. "Climate analogues suggest limited potential for intensification of production on current croplands under climate change." *Nature Communications*.

66 Davis, K.F. et al. 2017. "Increased food production and reduced water use through optimized crop distribution." *Nature Geoscience* 10: 919–24.

67 Minoli, S. et al. 2022. "Global crop yields can be lifted by timely adaptation of growing periods to climate change." *Nature Communications* 13: 7079.

68 Pontzer, H. et al. 2021. "Daily energy expenditure through the human life course." *Science* 373: 808–12.

69 Vollset, E. et al. 2020. "Fertility, mortality, migration, and population scenarios for 195 countries and territories from 2017 to 2100: A forecasting analysis for the Global Burden of Disease Study." *Lancet* 396: 1285–306.

Index

Africa: agriculture, 43, 74, 238; diet, 17, 34, 39, 67, 89, 131; environmental issues, 119, 123; food fortification, 142; food production as share of economy, 101; food supply, 14–15, 127, 144–6, 148, 149; food waste, 179; future agriculture and food supply, 176, 194–8; nutrient deficiencies, 140, 141, 142, 234; political stability, 145, 197–8; population growth projections and trends, 148, 201; soil fertility, 145; *see also individual countries by name*

agriculture: adapting crops to the environment, 175, 199; alternatives, 15–24; crop domestication, 29, 30, 31; crop selection, 29–47; definitions, 103; dependence on other industries, 103–7; disadvantages, 47–52; draft animals, 83; energy costs, 109–15; and the environment, 47–8, 116–17, 133–4; fallowing, 154–5; in the future, 149–71, 173–202; genetically modified crops, 158, 161–3; importance, 11–15, 24–7, 53, 206; increasing efficiency, 55–75, 144–6; mechanization and machinery, 46, 83, 84, 105–7, 115; monocropping, 173; multicropping and rotation, 154–7, 174, 175; organic farming, 2–3, 151–8; origins, 12–14; permaculture, 158–61; polycultures, 160–1; precision (high-tech) agriculture, 175–7; relationship to foraging, 11, 50; seed production, 106, 107, 156; as share of the economy, 101–24; *see also* animal farming and domestication

air pollution, 118–19, 120, 213
alcohol, 41, 72, 138, 198
Algeria, 206
allergies and intolerances: lactose, 37–8
almonds, 216
alpacas, 77
Amazon, 120–1
Americas: diet, 34, 39; draft animals, 83; *see also individual countries and regions by name*
amino acids, 37–8
Andes, 44–5
anemia, 140–1
Angola, 44, 198
animal domestication and farming: alternatives, 15–24; dependence on other industries, 106; draft animals, 83, 104; energy costs, 109–15; and the environment, 91–4, 116, 117; in the future, 149, 186–90, 193–4; headcount of

animal domestication and
    farming – *cont'd*:
    domesticated animals, 78–80;
    importance, 11–15, 24–5;
    intensive farming, 85–6; as share
    of the economy, 101–24; species
    choice for dairy farming, 83–4;
    species choice for draft purposes,
    83; species choice for meat,
    77–82; waste management, 106,
    122–3, 224; *see also individual
    animals by name*; dairy farming
    and products; meat
animal feed: cost, 107; for draft
    animals, 84; and the
    environment, 91–2, 186, 189–90;
    feeding efficiencies, 58–75, 80–2,
    86–9, 91–2, 95–8; in the future,
    149; and intensive farming, 85;
    oil cakes, 32, 66, 189, 245;
    phytomass as, 55, 67; types, 189
antibiotics, 168
antioxidants, 134–5
apples, 36, 72, 179; juice, 72
apricots, 30
aquaculture, 94–8, 106, 113, 149
Argentina, 84, 154, 196
arsenic, 122
artichokes, 134
Asia: agriculture, 154, 155; diet, 131;
    environmental costs of food
    production, 119; fish farming, 94,
    95; food production as share of
    economy, 101; food supply, 127,
    145–6; food waste, 179; future
    food supply, 199; nutrient
    deficiencies, 140, 141; population
    growth projections and trends,
    148, 201; *see also individual countries
    by name*

asses, 79
Australia, 71, 84, 97, 137, 149–50
Austria, 75
avocado, 16

bagasse, 32, 209
bananas, 30, *31*, 72
Bangladesh, 126, 128
barley: annual production, 25–6; and
    beer, 41; domestication process,
    29, 30; as food staple, 39;
    maturation period, 45–6;
    whole-grain consumption, 43;
    see also grains
beans: digestibility, 128;
    domestication process, 29, 33,
    34; as food staple, 39, 131, 132;
    see also soybeans
bears, 79
beef *see* cattle and beef
beer, 41, 72
beets, 45, 134
Benedict, Francis, 58
Benson, Andrew, 63
Binford, Lewis, 12
biofuels, 32, 150, 154, 209
biomass, 22, 23
biribá, 30–1, 208–9
bison, 20
black beans, 132
bodies, human: and energy efficiency,
    58; weight and mass, 9, 22;
    weight loss and dietary
    supplements, 233
Bolivia, 44–5
Brazil: agriculture, 32, 47, 68, 75,
    120–1, 154, 159, 208, 209; beef
    production, 92, 120–1; biofuels,
    150; diet, 44, 132; environmental
    issues, 120–1; food supply, 126,

128, 144; future food supply, 199;
    obesity, 118
bread, 34, 41, 51–2, 104, 179
broccoli, 45, 134
Brown, Lester, 191
Buck, John Lossing, 39–40
buffalo, 77, 79, 94
Bulgaria, 123, 182
Burkina Faso, 141
Burundi, 198

cabbages, 36, 134, 231
cadmium, 122
CAFOs *see* concentrated animal
    farming operations
calories, 15–16, 36, 126
Calvin, Melvin, 63
camels, 77, 79
Canada: agriculture, 45, 60–1, 66, 73;
    beef production, 84, 92; diet, 133;
    diet and health, 51; energy cost
    of food production and supply,
    109; food fortification, 139; food
    supply, 144; food waste, 179–80;
    life expectancy, 137; *see also*
    Americas; North America
cannibalism, 1
canola *see* rapeseed
car industry, 168
carbohydrates: diets high in, 130–1;
    diets low in, 89, 134–5; in grains,
    36; and health, 136; in hunted
    meat, 19–21; as ideal percentage
    of diet, 16, 127; nutritional value,
    15–16, 25–6
carbon dioxide: and decrease of
    organic matter, 174; and the
    environment, 117; and food
    supply, 198–9, 247; and
    genetically modified crops,
162–3; and photosynthesis, 55,
    58, 61–2, 70; soil storage, 175;
    and water use efficiency, 72–3
cardiovascular disease, 84, 135–6, 138,
    139
carp, 94, 95
carrots, 140
cassava, 29, 32, 38, 44
catastrophism, 1, 191–3
catfish, 95
Cathcart, Edward, 58
cattle and beef: beef eating and health,
    84; beef produced by grazing,
    91–3; beef's popularity, 84;
    biomass and diet, 21–2;
    domestication, 77, 80, 82; eating
    better, 188–90; eating less, 187–8;
    as draft animals, 83; and the
    environment, 91–4, 120–1,
    188–90; feeding efficiency, 80, 82,
    86–9, 91–2, 220; headcount of
    domesticated, 78–9; intensive
    farming, 85; time spent
    grazing, 23
cauliflower, 36, 134
Central African Republic, 127, 145
cereals: as animal feed, 85, 91, 95, 200;
    by-products, 41; contamination,
    122; and fallowing, 155; as food
    staples, 26, 32, 83, 138; future
    uses, 200; genetically modified,
    161–2; global harvest size, 32;
    imported and exported, 145, 149;
    and land use, 25; nutritional
    value, 36, 37–41; organic yield,
    153; photosynthetic efficiency,
    59–63, 64–6; polycultures, 160–1;
    and rotation, 157; transportation,
    114; wastage statistics, 179; water
    use efficiency, 71–2, 92, 134;

cereals – *cont'd*:
see also individual cereals by name;
grains
cheese, 140
cherries, 30
chicken: cultured, 163; digestibility, 128; and the environment, 94, 174; feeding efficiency, 87–9, *87*; in the future, 190; headcount of domesticated, 79; intensive farming, 85, 86; nutritional value, 38; popularity, 146; waste management, 224
chickpeas, 29, 33, 34, 38
Childe, Gordon, 11
children: childhood malnutrition and economic growth, 146; nutrient deficiencies, 140, 141, 142
Chile, 152
chimpanzees, 5–7, 6, 9, 18, 53
China: corn growing, 67; diet, 34, 39–40, 41–2, 43, 89–90, 131, 193; diet and health, 138; economic growth, 101; energy cost of food production and supply, 109, 112, 114–15, 225; environmental issues, 121–2, 193–4; fertilizer use, 74, 75; fish farming, 95; food production and employment, 108; food storage, 112; food supply, 126, 127, 128, 145–6, 148; food waste, 181–2; future food supply, 191–2, 193–4, 199, 200; grain growing, 45–6, 71; grain storage and supply, 46, 47; household food expenditure, 109, 185; legume growing, 46; meat farming, 81, 218; population change projections, 148; rapeseed growing, 66; rice growing, 43, 66–7, 154, 160, 194; saving face, 181; soil contamination, 122; sugarcane growing, 208; traditional cropping and population density, 14
Chinese gooseberries, 30
cholesterol, 135–6
climate: and origins of agriculture, 12; prehistoric, 10
climate change: crop adaptation, 175, 199; and cultured meat, 168; and food production, 47–8, 93–4, 117, 118, 174, 175; and future food production and supply, 198–201
coffee, 72, 92
cogeneration, 73
concentrated animal farming operations (CAFOs), 84, 85–6
Congo, 145, 198
conifers, 30
cooking *see* food preparation
corn: as animal feed, 84, 85, 89; annual production, 25–6; and biofuels, 32, 150; cooking oil from, 41; crop yields, 144, 195–6, 198, 199; domestication process, 29, 30, 34; fertilizer use, 74; genetically modified, 161; maturation period, 46; photosynthetic efficiency, 63–4, 67–8; seed production, 156
courgettes *see* zucchini
cowpeas, 29, 39, 196
cows *see* cattle and beef
Croatia, 233
crustaceans, 94–5, 96, 190
cucumber, 45

dairy farming and products, 83–4, 94, 140, 149, 179; *see also* eggs; milk
dates, 17, 72

Davis, William, 50–1
deer and venison, 20–1
deforestation, 120–1
Denmark, 138, 187
Diamond, Jared, 48, 49
diet: and health, 125–46
Dobzhansky, Theodore, 98
donkeys, 77
DRC *see* Congo
drones, 175–6
ducks, 79, 218–19

eating out *see* restaurants
economy, global: food production's share, 101–24; services' share, 101–2; smartphones' share, 102
Eden, Frederick Morton, 83
Edison, Thomas, 57
eels, 96
eggs: cultured, 163; as desirable sources of protein, 188; digestibility, 128; and the environment, 94; expiry dates and safety, 177; in the future, 190; morality of eating, 129; nutritional value, 37, 38
Egypt, 34, 81, 206
Egypt, ancient, 13–14, 34, 35, 46
electricity: and energy efficiency, 57–8
employment and labor: and manure application, 158, 238; percentage of market employed in food supply, 108; and perennial crops, 160; in pre-mechanized agriculture, 46; in Roman bakeries and cookshops, 104; and vegan food supply, 133
energy cost: of cultured meat, 165, 168; of food preparation and refrigeration, 114–15, 225; of food production and supply, 109–15; of food transportation, 114; and permaculture, 159; of wasted food, 177
energy efficiency: of fish, 95–8; in general, 56–8; of meat, 80–2, 86–9, 91–2; of plants, 58–75, 134
Engel, Ernst, 108
engines: and energy efficiency, 56–7
England: diet, 83; population density and agriculture, 14; *see also* UK
environment: and agriculture, 47–8, 116–17, 133–4; air pollution, 118–19, 120, 213; crop adaptation, 175, 199; deforestation, 120–1; and fish farming, 97–8; and food production, 116–23; insulation, 178, 242; lessening food supply's impacts, 147–71, 173–90; and meat and dairy farming, 85–6, 91–4, 116, 117, 120–1, 130–1, 174, 176, 186–90; and nuts, 134; recycling, 178, 242; soil pollution and erosion, 118–19, 120, 121–2, 158–9, 173; water pollution and depletion, 116–17, 118–19, 120, 121, 173–4; *see also* climate change
Eritrea, 198
ethanol, 32, 150, 209
Ethiopia: food staples, 33; food supply, 15, 126, 128, 144; nutrient deficiencies, 143; political stability, 145, 198
Europe and European Union: agriculture, 45–6, 66, 67, 74, 75, 154, 175; animals per hectare, 122–3; diet, 39, 40, 131, 146, 188; diet and health, 51–2; draft animals, 83; energy cost of food preparation, 114; fish farming, 94;

Europe and European Union – *cont'd*:
food fortification, 139; food
waste, 179; future food supply
and the environment, 149–50;
genetically modified crops,
161–2; household food
expenditure, 109, 185; life
expectancy, 137; nutrient
deficiencies, 141; population size
projections, 148; *see also individual
countries by name*

fallowing, 154–5
farming *see* agriculture; animal
domestication and farming; fish:
farming
fats: global supply, 128; in grains, 36;
and health, 135–6, 138; in hunted
meat, 19–21; as ideal percentage
of diet, 16, 127; modern
preferences, 146; nutritional
value, 16, 41–2
feeding efficiency *see* energy efficiency
fertilizers: energy costs and efficiency,
69, 105, 111, 112; and the
environment, 116–17, 122; and
future food supply, 195–6;
manure as, 104, 157–8, 175;
nitrogenous, 73–5, 116–17; and
organic farming, 151–8; and
permaculture, 159; US annual
expenditure, 107
fibre, 134–5
fibrous crops, 31
figs, 5, 17, 72, 206
financial crisis (2008), 192, 245
Finland, 232
fish: cultured, 163; farming, 94–8, 106,
113, 149, 188, 190; fish oil, 134;
and health, 138, 140; trophic
levels, 79; wastage statistics, 179;
see also individual fish by name
fishing: by early humans, 8; and
hunter-gatherer population
densities, 9
flax, 33
flour: extraction rates, 43, 66;
fortification, 40, 139, 211; and
health, 50; no-gluten, 34; soya,
38; uses, 41
food expenditure, household, 108–9,
185
food preparation, energy costs of,
114–15, 225
food pricing, 185, 194
food production and supply: and aging,
200; daily statistics, 2; diet and
income, 128; and the economy,
110–24; and employment, 108;
energy cost, 109–15; and the
environment, 116–23, 147–71,
173–90, 198–201; in the future,
147–71, 173–202; global variation,
14–15; inadequate, 118, 126–7;
increasing, 144–6, 147;
inequalities, 191–8; modern
dietary trends, 146; storage and
transportation, 40–1, 42–3, 46–7,
112, 114, 182–3; variety, 184–5
food safety, 122, 177
food staples, 31–47
food storage and transportation, 40–1,
42–3, 46–7, 112, 114, 182–3
food waste: and affluence, 181; causes,
14–15, 90, 177; daily, 2; as
percentage of US total waste,
112; percentages by types of
food, 179; reducing, 176–86
foraging: modern idealization, 48–50;
nutritional value of foraged

items, 16; and population
density, 9, 13; prehistoric, 7–9;
relationship to agriculture, 11, 50
foxes, 79
Framingham Heart Study, 135
France: diet, 90, 131, 138, 187–8, 233;
diet and health, 135–6; draft
animals, 83; energy cost of food
production and supply, 109; food
production as share of economy,
101; food supply, 144, 148; wine
harvests and climate change, 198
Francione, Gary, 98
fruit: advantages of domesticated, 30;
and chimpanzee diet, 5, 7; as
food staple, 39–40; fruit eating
and income, 128; fruit eating and
population densities, 7; and
health, 138; juices, 72; most
important species, 206;
nutritional value, 7, 16–17, 36,
134; wastage statistics, 179, 180;
water use efficiency, 72; *see also
individual fruits by name*
fuel: agricultural use, 106, 107;
biofuels, 32, 150, 154, 209; *see also
energy cost; energy efficiency*

GDP *see* gross domestic product
geese, 78, 79, 218
genetically modified crops, 158, 161–3
Germany: diet, 131, 138, 187; diet and
health, 51, 52; food production as
share of economy, 101; food
supply, 148; household food
expenditure, 109; organic
farming, 153–4
Ghana, 44, 241–2
global positioning systems (GPS), 175
global warming *see* climate change

glucose, 59–60, 61–2
gluten, 34
goats, 77, 78–9, 81
Göbekli Tepe, 50
gorillas, 17–18, 49, 53
grain legumes: alternatives, 132–3;
digestibility, 132, 230;
domestication process, 38; effects
of increasing consumption,
131–2; and fertilization, 151–6,
174; maturation period, 46;
nutritional value, 37–41, 131;
popularity, 146; water use
efficiency, 72; *see also individual
grain legumes by name*
grains: advantages as foodstuffs, 13,
40–1; as animal feed, 85, 91, 95;
annual production, 25–6;
by-products, 41; disadvantages as
crop, 47–8, 50; domestication
process, 30, 33; fertilizer use, 73,
74; as food staple, 32; fortified
flour, 40, 139, 211; future uses,
200; genetically modified, 161–2;
and health, 50–2; heavy metal
contamination, 122; maturation
period, 45–6; milling, 43, 47, 66;
nutritional value, 35–41, 66;
perennial crops, 159–60;
photosynthetic efficiency, 59–63,
64–6; polycultures, 160–1;
post-harvest losses, 41–3; societal
impacts, 45–6; storage and
transportation, 40–1, 42–3, 46–7,
114; wastage statistics, 179; water
use efficiency, 71–2, 92, 134; *see
also individual grains by name;
cereals*
grapes, 72, 198
Greece, 232, 233

greenhouse gases, 117, 118, 174; *see also* carbon dioxide; methane
gross domestic product (GDP), 101
guinea pigs, 79, 80–1

Hadza tribe, 8–9
hares, 21, 78, 79
Hatch, Hal, 63
health: and diet, 50–2, 84, 125–46, 135–6, 138, 140; dietary supplements, 139–43, 233; food fortification, 139–43; health costs of food supply, 118, 244; improving, 139–46; life expectancy, 51–2, 136–8, 244; nutrient deficiencies, 140–3; obesity, 14–15, 118, 229; *see also* nutrition
herbs, 31
herring, 79
Heyden, Pieter van der: engravings by, 65
Hinduism, 81, 83
hominins, 7–10
horses, 79, 83, 84, 187
house insulation, 178
Hu, Frank, 136
Hughes, Huw, 166–7
humans, early, 7–10
hunter-gatherers, 7–10; *see also* foraging
hunting, 6–10, 18–21

Iceland, 90
immigration, 148, 201
India: agriculture, 33, 43, 46, 74, 75, 208; diet, 39, 40, 44, 83, 89–90, 128, 131, 132; economic growth, 101; energy cost of food preparation, 114–15; environmental issues, 121; food fortification, 142; food supply, 126, 127, 128, 145, 148–9; household food expenditure, 109; nutrient deficiencies, 140; population growth projections, 148
intoxicants, 31, 41, 72, 138, 198
Inuit, 39
iodine deficiency, 141–2
Iran, 77, 206
Iraq, 77
Ireland, 75
iron deficiency, 140–1
irrigation *see* water; water use efficiency
Islam, 81
Italy: agriculture, 75; diet, 233; diet and health, 51, 52; fish farming, 97; food supply, 148; life expectancy, 51, 52, 137; rabbit farming, 81

Jackson, Wes, 159
Japan: agriculture, 45, 66–7; diet, 37–8, 42, 43, 83–4, 131, 133, 188; diet and health, 51, 137–8; energy cost of food production and supply, 109; fish farming, 97; food production as share of economy, 101; food supply, 126, 146, 148, 200; food supply and health, 244; future food supply and the environment, 149–50; household food expenditure, 185; population change projections, 148
Judaism, 81

kale, 134
Kazakhstan, 140
Kenya, 234
Keys, Ancel, 135

kilocalories, 15–16, 36, 126
Kok, Björn, 96
Kortschak, Hugo, 63

labor *see* employment and labor
lactose intolerance, 37–8
lamb *see* sheep and lamb
land use, 116, 117, 120–1, 153–4, 193–4
Latin America: agriculture, 154; diet, 131; future food supply, 199; nutrient deficiencies, 140, 141; *see also individual countries by name*; Americas
lead, 122
legumes *see* grain legumes
Lehman Brothers, 245
lentils, 3, 34, 39, 131, 132
lettuce, 36, 45, 134
Liberia, 198
life expectancy, 51–2, 136–8, 244
light bulbs, 57
lions, 79
liver, 140
llamas, 77
Loomis, Robert, 161

Macedonia, 34
mackerel, 79
Madagascar, 127
Mali, 141, 198
Malthus, Thomas Robert, 1
mammoths, 8, 18–19
Manila, 10
Manitoba, 60–1, 64, 66, 73
manure, 104, 106, 122–3, 157–8, 175, 224
Marlowe, Frank, 8–9
meat: annual global production, 168; and chimpanzee diet, 6–7; choice of species to eat, 80–2; cultured, 163–71, *164*, *166*; digestibility, 128; and early human diet, 7–10; effects of increasing consumption, 130–1, 230; and the environment, 85–6, 91–4, 116, 117, 130–1, 174, 176, 186–90; expiry dates and safety, 177; feeding efficiency, 2, 80–2, 86–9, 91–2; in the future, 187–90, 193–4; and health, 84, 138, 140; intensive farming, 85–6; meat eating and income, 128; meat produced by grazing, 91–3; morality of eating, 85–6, 98–9; most popular, 84, 146; nutritional value, 16, 18–21, 38, 82; Paleolithic diet, 89, 130–1; as percentage of global food energy and dietary protein, 130–1; plant-based, 132–3, 169; and population growth, 11–12; recent consumption trends, 133, 174; wastage statistics, 179, 180; *see also individual animals by name*; animal domestication and farming
meatless meat, 132–3, 169
Mediterranean diet, 135–6, 233
megaherbivores, 8, 18–19
mercury, 122
methane, 93–4, 117, 174
Mexico: agriculture, 34, 39; diet, 67; food supply, 128; obesity, 118
mice, 80
Middle East, 123
migratory societies, 24
milk: cultured, 163; digestibility, 38, 128; and the environment, 94; as food staple, 83–4, 89, 146; in the future, 149; lactose intolerance,

milk – *cont'd*:
  37–8; nutritional value, 37; and veganism, 129
millet, 29, 34, 43, 63–4
miso, 132
Monbiot, George, 1, 192
Morocco, 206
Mozambique, 145, 198
mules, 79
mushrooms, 55

Netherlands: agriculture, 14, 72–3, 134; animal farming, 122–3, 149; cultured meat industry, 163, 169
Newcomen, Thomas, 56
Niger, 198
Nigeria: agriculture, 43; diet, 44; food supply, 126, 128, 144–5; political stability, 198
nitrogen: and genetically modified crops, 161–2; nitrogenous fertilizers, 73–5, 116–17; and organic farming, 151–8; and photosynthesis, 60, 70; and rice growing, 195
nitrogen use efficiency (NUE), 74–5
nitrous oxide, 117, 174
noodles, 41, 132
North America: agriculture, 45–6, 66; food waste, 179–80; future food supply and the environment, 149–50; population growth projections, 148; precision (high-tech) agriculture, 175; *see also* Canada; USA
North Korea, 81
Norway, 52, 75
nutrition: aging's effects on requirements, 200; and carbohydrates, 15–16, 25–6; deficiencies, 140–3; dietary supplements, 139–43, 233; extreme diets, 89, 98–9, 129–36; and fats, 16, 41–2; food fortification, 139–43; and fruit, 7, 16–17, 36, 134; global undernourishment statistics, 118, 126–7; and grains and legumes, 35–41, 66; human requirements, 15–16, 35–41, 200; improving, 139–46; kilocalories needed daily, 15–16, 36, 126; and meat, 18–21, 38, 82; overview, 125–46; and plants, 17–18, 21–2, 25–6; and proteins, 15–16; real food, 136; science of, 125–6; and wood, 23–4
nuts, 16, 72, 134, 216; *see also* peanuts

oats, 34, 39, 45–6
obesity, 14–15, 118, 229
oil cakes, 32, 66, 189, 245
oils *see* plant oils
olive oil, 72
oranges, 30, 36; juice, 72
organic farming, 2–3, 151–8
Ormesson, Jean d', 49
oxen, 83
oxygen, 52

packaging, 182, 183
Pakistan, 126
Paleolithic diet, 89, 130–1
pasta, 38, 41, 51–2, 60
peaches, 30, 72, 131
peanuts: cooking oil from, 41, 72; domestication process, 29; as food staple, 39, 145; water use efficiency, 72
pears, 72
peas, 33, 34, 39, 131

peppers, 72–3
permaculture, 158–61
Peru, 44–5
pesticides, 106, 107, 151–2, 155, 158
pharmaceutical industry, 167–8
phosphorus: agricultural needs, 71, 73; in fertilizers, 105, 152, 196; foods rich in, 40, 69; and high yields, 157; and photosynthesis, 55; and rice growing, 195
photorespiration, 62–3
photosynthetic efficiency, 55–6, 58–75, 161–3
photovoltaic cells, 57–8
pigs and pork: consumption statistics, 81–2; diet, 79; domestication, 77, 81–2; eating less, 187; and the environment, 94; feeding efficiency, 82, 87–8; in the future, 190; headcount of domesticated, 79; and health, 138; intensive farming, 85, 86; muscle and fat share, 219; nutritional value, 38, 82
pine nuts, 30
pineapples, 72
plant oils, 32, 66, 72, 135–6, 179
plants: agricultural crop selection, 29–47; digestibility by humans, 18, 21–3; domestication process, 29, 30–1; edible phytomass, 64–6; and the environment, 133–4; increasing efficiency, 55–75; nutritional value, 17–18, 21–2, 25–6; photosynthetic efficiency, 55–6, 58–64; seed production, 106, 107, 156; total number identified, 29; *see individual plants and plant types by name*
plastic, 178, 242

plums, 30, 72
polycultures, 160–1
population: agriculture and growth, 11–15, 206; growth projections and trends, 148, 201; prehistoric, 10; Roman, 11–12
population densities: and agriculture, 12, 13–15; and fruit eating, 7; hunter-gatherers, 8–10
pork *see* pigs and pork
potassium: agricultural needs, 73; in fertilizers, 105, 152; foods rich in, 69; and high yields, 157; and photosynthesis, 55
potatoes: crop yields, 196; domestication process, 29; as food staple, 32, 39–40; maturation period, 46; nutritional value, 16; photosynthetic efficiency, 66; pros and cons as crop, 44–5; wastage statistics, 179
potatoes, sweet, 32, 46, 140
poultry *see individual birds by name*
precipitation *see* rainfall
proteins: global supply, 128; in grains and legumes, 36, 37–40; in hunted meat, 19–21; as ideal percentage of diet, 16, 127–8; nutritional value, 15–16; sources, 128, 130–3
pulses, 128, 131–2; *see also individual pulses by name*; grain legumes

quinoa, 29

rabbits: headcount of domesticated, 79; as meat source, 80–1; nutritional value, 19, 21
rainfall, 121, 123

rapeseed (canola), 29, 66, 161
rats, 80
recycling, 178, 242
refrigeration, 115; intelligent refrigerators, 183
restaurants: and food waste, 181, 183; Roman, 104, 223; US expenditure, 111
rice: adapting to climate change, 175; by-products, 41; crop yields, 144; domestication process, 29, 30; double-cropping, 154; fertilizer use, 74, 195; as food staple, 34, 39–40; in the future, 194, 199; maturation period, 45, 46; milling, 43; nutritional value, 37, 39, 69; as percentage of global food energy, 29; perennial crops, 159–60; photosynthetic efficiency, 59–67, 69; post-harvest losses, 43; value of annual global harvest, 102; *see also* grains
Romania, 109
Romans: bakeries and cookshops, 104, 223; diet, 34, 46–7, 80, 83; population, 11–12; population density and agriculture, 14; transportation of foodstuffs, 104
root crops, 179; *see also individual roots by name*
Rubisco, 59, 62, 162, 199
Russia: agriculture, 159; food supply, 149; rabbit farming, 81; wheat exports, 47; *see also* Soviet Union
Russo-Ukrainian War, 149, 185
Rwanda, 198
rye, 34, 39, 45–6

salmon, 95, 96–7, 97–8
salt, 139, 142

Scott, James C., 50
seafood, 180; *see also* crustaceans; fish
seals, 19–20
seed production, 106, 107, 156
seitan, 41
Senegal, 6
sharks, 79
sheep and lamb, 77, 80, 81, 187
Sierra Leone, 198
Slack, Roger, 63
Slovakia, 182
smartphones, 102
Smil, Vaclav: other books about food, 3
smoking, 138
soil: environmental management and monitoring, 175; pollution and erosion, 118–19, 120, 121–2, 158–9, 173; rooting depths, 195–6
solar energy, 57–8; and photosynthesis, 58–61, 64
Somalia, 198
Somerville, Chris, 162
sorghum: domestication process, 29; maturation period, 46; nutritional value, 38, 39; photosynthetic efficiency, 63–4
South Africa, 234
South Korea, 37–8, 146, 148
South Sudan, 198
Soviet Union: agriculture, 75; labour camp diets, 210; *see also* Russia
soybeans: as animal feed, 84, 85, 89; cooking oil from, 41, 72; domestication process, 29, 30; and the environment, 120–1; as food staple, 34, 39, 131; maturation period, 46; miso, 132; nutritional value, 38, 39; soy

sauce, 41, 132; soya flour, 38; tofu, 39, 41, 132–3; transportation, 47; water use efficiency, 72
Spain, 51, 52, 75, 134, 137–8
spices, 31
squashes, 140
steam power, 56–7
Stefansson, Vilhjalmur, 19
straw, 64–5
Sudan, 145, 198
sugar, 128
sugarcane: and biofuels, 32, 150; domestication process, 29; as food staple, 39–40; nutritional value, 25–6, 131, 230; perennial crops, 159; photosynthetic efficiency, 63–4, 68–9
sunflowers, 30
Sweden, 137
sweetcorn, 67–9
Switzerland, 52

Tanzania, 6, 8–9
termites, 6, 7, 23–4
Thailand, 43, 208
tilapia, 95
tofu, 39, 41, 132–3
tomatoes, 36, 72–3, 134
trophic levels, *78*, 79, 89–91, *90*, 95–6
trout, 95, 96–7
tubers *see* cassava; potatoes
tuna, 79, 97
Turkey, 34, 50, 77, 206
turkeys, 79, 86, 218

Uganda, 234
UK: agriculture, 75; animals per hectare, 122; diet, 40, 53; food production as share of economy, 101; food supply, 126; food waste, 179; population density and agriculture, 14
Ukraine, 149, 185
ultra-processed foods, 136
USA: agricultural machinery market, 106–7; animals per hectare, 122; beef production, 84, 85, 86, 92, 93; biofuels, 150; corn growing, 32, 67–8, 155, 156, 161, 195, 209; diet, 88, 133, 146, 187, 193; dietary guidelines, 135–6; dieting, 229; energy cost of food production and supply, 109, 110, 111–12, 114; environmental issues, 118–19, 123; fish farming, 97; food fortification, 139; food imports, 111–12; food production and employment, 108; food production as share of economy, 101, 107–8, 119; food supply, 126, 144; food variety, 184–5; food waste, 112, 177, 179–80, 183; genetically modified crops, 161; grain growing, 71, 159; household food expenditure, 108, 111, 185; intensive farming, 85, 86–9; life expectancy, 137, 138; methane sources, 93; milk production, 83; multicropping, 154, 155; nutrient deficiencies, 141; obesity, 229; restaurant portion size, 183; service economy, 101; soybean exports, 47; sugarcane growing, 208; turkey farming, 218; vegetable growing, 133–4; vegetarians as percentage of population, 218; *see also* Americas; North America

vaccination, 139
Van Nieuwkoop, Martien, 117–18, 119
veganism: feasibility of increasing, 131–4, 170–1, 186; morality of, 98–9; permitted foods, 129
vegetables: and the environment, 133–4, 186; as food staple, 39–40; maturation periods, 45–6, 134; multicropping, 154; nutritional value, 36; wastage statistics, 179, 180; water use efficiency, 72; *see also individual vegetables by name*
vegetarianism: morality of, 98–9; permitted foods, 129; protein intake, 38; vegetarians as percentage of population, 77, 218
venison *see* deer and venison
vinegar, 41
vitamin deficiency, 140, 234

walnuts, 216
walruses, 19–20
water: and fallowing, 155; food production's share of global resources, 116, 120; and food waste, 177; global changes in water cycle, 198, 245; and nuts, 134; and permaculture, 158–9, 161; and photosynthesis, 58, 69–73; pollution and depletion, 116–17, 118–19, 120, 121, 173–4, 194; rainfall, 121, 123; and vegetable growing, 133–4
water buffalo, 77, 79
water use efficiency, 71–3, 91–2
watermelons, 72
Watt, James, 57
weight *see* bodies, human
whales, 19–20

wheat: as animal feed, 89; annual production, 25–6; crop yields, 199; digestibility, 128; domestication process, 29, 30; double-cropping, 154; fertilizer use, 73, 74; as food staple, 33–4, 39–40; fortified flour, 40, 139, 211; and health, 50–2; maturation period, 45–6; milling, 43; nutritional value, 36, 38, 39, 66; as percentage of global food energy, 29; perennial crops, 159; photosynthetic efficiency, 59–63, 64–6; post-harvest losses, 42–3; seitan, 41; storage and transportation, 46–7; value of annual global harvest, 102; water use efficiency, 71; *see also* cereals; grains
wine, 72, 138, 198
women: and cardiovascular disease, 138; diet, 188; life expectancy, 137, 232; nutrient deficiencies, 140, 141, 142; protein requirements, 128; and vegetarianism, 218
wood, 23–4
Wood, Paul, 166

yaks, 77
yams, 32
yogurt, 55, 84, 146, 177, 184–5

Zambia, 141
Zangwill, Nick, 98
Zimbabwe, 127
zinc: deficiency, 140, 142–3, 246; importance, 40, 55
zucchini, 45

100 YEARS of PUBLISHING

Harold K. Guinzburg and George S. Oppenheimer founded Viking in 1925 with the intention of publishing books "with some claim to permanent importance rather than ephemeral popular interest." After merging with B. W. Huebsch, a small publisher with a distinguished catalog, Viking enjoyed almost fifty years of literary and commercial success before merging with Penguin Books in 1975.

Now an imprint of Penguin Random House, Viking specializes in bringing extraordinary works of fiction and nonfiction to a vast readership. In 2025, we celebrate one hundred years of excellence in publishing. Our centennial colophon will feature the original logo for Viking, created by the renowned American illustrator Rockwell Kent: a Viking ship that evokes enterprise, adventure, and exploration, ideas that inspired the imprint's name at its founding and continue to inspire us.

For more information on Viking's history, authors, and books, please visit penguin.com/viking.